D0597179

Knowledge Management
and
Its Integrative Elements

Knowledge Management

and
Its Integrative Elements

Edited by

Jay Liebowitz
Professor of Management Science
George Washington University

Lyle C. Wilcox
Vice Chancellor for Technology
State College and University Systems of West Virginia

CRC Press
Boca Raton   Boston   London   New York   Washington, D.C.

Library of Congress Cataloging-in-Publication Data

Knowledge management and its integrative elements / edited by Jay
   Liebowitz, Lyle C. Wilcox.
      p.   cm.
   Includes bibliographical references and index.
   ISBN 0-8493-3116-1 (alk. paper)
    1. Information resources management. 2. Expert systems (Computer
science) I. Liebowitz, Jay. II. Wilcox, Lyle Chester.
  T58.64.K65 1997
  658.4′038′0285633--dc21
                        97-305
                        CIP

No claim to original U.S. Government works
International Standard Book Number 0-8493-3116-1
Library of Congress Card Number 97-305
Printed in the United States of America   2 3 4 5 6 7 8 9 0
Printed on acid-free paper

Dedication

To our wonderful students over the years and our supportive families — Janet, Jason, and Kenny Liebowitz, and Pat, Michael, Beth, Vicki, and Laura Wilcox; and to Raymond J. Wilcox, whose guidance and encouragement leads people to extraordinary achievement.

Preface

As we continue to move into the Knowledge Age, the ability of organizations to manage, store, value, and distribute knowledge will be paramount. This capability is known as "knowledge management." Information with wisdom may be considered "knowledge," and organizations have found that "knowledge" is their most competitive weapon and edge against others.

A critical component of knowledge management is the development, application, deployment, and sharing of "knowledge-based systems." Knowledge-based systems are an ideal technology for preserving knowledge within the organization and building the corporate memory of the firm. Knowledge-based systems also play a powerful role in integrating knowledge across departments, disciplines, and organizations. Activities such as concurrent engineering, corporate synergy, and dynamic process development are all products of the emergence of knowledge-based thinking.

The teaching and learning process is being dramatically influenced by the instruments that have been developed to collect, repackage, and communicate complicated ideas and issues. Access to courses and courseware is being routinely provided through the powerful and rapidly advancing technologies associated with the intranet concepts. Information and knowledge take form and substance at common workplaces where ideas are born and develop into new and higher orders of knowledge. In this context, knowledge is a commodity, constantly developing and building.

In this book, we have demonstrated the breadth and depth of the work being done in knowledge-based systems. From classroom to corporate strategic planning, the concepts of KBS are forcing redefinitions of management principles along with totally new constructs of organizational design. Clearly the successful, whether individuals or organizations, will be those that have cast aside old paradigms and replaced them with new visions created by the learner-based society.

The chapters of this book illustrate the powers of knowledge-based system thinking in the applications environment, while simultaneously introducing the reader to the intricacies and complexities of the computers, networks, inference engines, and other enabling technologies that furnish the user with such powerful management and research tools.

This book should be required reading for senior management, information/knowledge technology professionals, educators, and those interested in becoming successful managers in organizations. Many of the leading authorities in the knowledge management field have contributed chapters in this book. We are very pleased to have their participation.

We acknowledge the participation of Ron Powers and his staff at CRC Press, George Washington University, Marshall University, the U.S. Army War College, and the West Virginia Higher Education Chancellor's Office for their continuing support.

As we close our effort on this book, we already look to the future in seeking out and presenting for our readers even more advanced and exciting contributions to the field of KBS from politics to polymers. We hope you will find this book to be as enjoyable for you as it was for us to compile.

Jay Liebowitz
Lyle C. Wilcox

Editors

Dr. Jay Liebowitz is Professor of Management Science at George Washington University in Washington, D.C. He recently held the Chair in Artificial Intelligence in the Center for Strategic Leadership at the U.S. Army War College. He is Editor-in-Chief of the international journal, *Expert Systems With Applications,* published by Pergamon Press/Elsevier. He is also Editor-in-Chief of a new international journal, *Failure and Lessons Learned in Information Technology Management,* published by Cognizant Communication Corporation in New York.

Dr. Liebowitz was selected as the Computer Educator of the Year 1996 by the International Association for Computer Information Systems. He is the Founder and Chair of The World Congress on Expert Systems, and has published 18 books and close to 200 articles.

Lyle C. Wilcox, Ph.D., is Vice Chancellor of the State College and University Systems of West Virginia, and has been appointed by the governor as the Chairman of the Science and Technology Advisory Council for West Virginia.

Dr. Wilcox has held tenured faculty appointments at Michigan State University, Clemson University (South Carolina), James Madison University (Harrisonburg, VA), and Marshall University (Huntington, WV).

Dr. Wilcox served as the Dean of Engineering at Clemson University for 8 years. He was the president of the University of Southern Colorado and following that was appointed to the position of Deputy Assistant Secretary in the Department of Energy. From 1986–1991, Dr. Wilcox held positions of Senior Vice President for Research at Telex Corporation and Vice President of Research for Purolator Corporation. He served as the founder and Provost of the College of Integrated Science and Technology at James Madison University and later became Senior Vice President and Provost at Marshall University.

Dr. Wilcox has been principal investigator for a number of research grants and projects in the areas of smart sensor development, modeling of industrial processes, pattern recognition, and biotechnologies. More recently, Dr. Wilcox has been an active contributor to the field of knowledge-based systems and expert systems applications. He led the efforts of both Telex Corporation and Purolator Corporation in expert systems, and carried this experience into the development of new undergraduate programs in knowledge-based systems in the academic environment. He has received the U.S. Department of Energy Distinguished Service Award, the Idaho National Engineering Laboratory Achievement Award, and has been elected to receive the Distinguished Service Award from his alma mater, Tri-State University. He has also received the National Science Foundation Faculty Award. He was given the Commendation for Outstanding Leadership by the Oklahoma Center for the Advancement of Science and Technology.

Dr. Wilcox has served as chairman of the board of the Oklahoma Independent Colleges Foundation, was appointed as a board member of the Oklahoma Pollution Control Coordinating Committee, was an executive board member for the Oklahoma Center for the Advancement of Science and Technology, has been a trustee of Tri-State University in Angola, IN, and served on the White House Committee assessing U.S. policy concerning scientific and technological manpower exchange between the U.S. and foreign countries.

He has held membership in the Institute of Electrical and Electronic Engineers, the American Management Society, and American Society of Mechanical Engineers, and is a member of the honorary societies Tau Sigma Eta and Phi Kappa Phi.

Dr. Wilcox has prepared papers for and made presentations to the National Academy of Sciences, the Applied Energy International Symposium on Reactor Development (Tokyo), the Brookings Institute, the University of Oklahoma, and the American Nuclear Society. His most recent interests are in the field of knowledge-based systems, where he has completed an invited paper in *Expert Systems with Applications,* has presented "Engineering for the 21st Century" at a national conference of the ASME, has written a manuscript on "The Concept of Professional Transitioning," and has published and presented on the subjects of integrated studies and knowledge-based systems at four regional and national meetings.

Contributors

Ali Bahrami
Center for Management and
  Technology
Rhode Island College
Providence, Rhode Island

Johan C.M. den Biggelaar
CIBIT
Utrecht, The Netherlands

Christopher J. Fox
College of Integrated Science and
  Technology
James Madison University
Harrisonburg, Virginia

George Hluck
U.S. Army War College
Center for Strategic Leadership
Science and Technology Division
Carlisle Barracks, Pennsylvania

Joe Marchal
College of Integrated Science and
  Technology
James Madison University
Harrisonburg, Virginia

Richard M. Roberds
College of Integrated Science and
  Technology
James Madison University
Harrisonburg, Virginia

William J. Barnett
Fluor Daniel
Greenville, South Carolina

Bruce O. Blagg
Transformation Institute
Tampa, Florida

Tom Galvin
U.S. Army War College
Center for Strategic Leadership
Science and Technology Division
Carlisle Barracks, Pennsylvania

Robert de Hoog
Department of Social Science
  Informatics
University of Amsterdam
Amsterdam, The Netherlands

Mohsen Modarres
Department of Management and
  Systems
Eastern Washington University
Spokane, Washington

Rob van der Spek
CIBIT
Utrecht, The Netherlands

André Spijkervet
CSC Computer Sciences
Amsterdam, The Netherlands

Jerry Benson
James Madison University
Harrisonburg, Virginia

W. Joseph Cochran
Transformation Institute
Tampa, Florida

Phillip Kevin Giles
U.S. Army War College
Center for Strategic Leadership
Science and Technology Division
Carlisle Barracks, Pennsylvania

Jay Liebowitz
Department of Management Science
The George Washington University
Washington, D.C.

Jack Presbury
Department of Psychology
James Madison University
Harrisonburg, Virginia

Arun Vedhanayagam
Transformation Institute
Tampa, Florida

Karl Wiig
Knowledge Research Institute
Arlington, Texas

Lyle C. Wilcox
Vice Chancellor for Technology
State College and University Systems of
  West Virginia
Charleston, West Virginia

Contents

1 Knowledge-Based Systems as an Integrating Process

Lyle C. Wilcox

CONTENTS

INTRODUCTION

The contributions of the authors in this book represent an impressive demonstration of the depth and breadth that the study of knowledge-based systems (KBS) has brought to the world of multidisciplinary problem solving. The diversity of this material illustrates the nature of the kinds of problems brought about by the dramatic economic and social changes occurring globally. Many contemporary issues of interest are not confined to a single discipline or professional classification. In the business arena, one observes a constant reference to re-engineering and reorganization, indicating dramatic shifts in not only what is done but how it is done. Education is feeling the backlash brought about by the resulting job and career changes. The entire teaching-learning structure, from curriculum and courseware to delivery, is the subject of intensive discussion and evaluation. This book illustrates how the knowledge domains of experts can be fused into powerful new methodologies for higher level thinking and complex problem-solving; the transition that is apparent to many of us is that we have moved from multidisciplinary studies to "integrated studies."

This chapter takes a look at the evolutionary process that seems to be spawning a new breed of scientist/manager/researcher/educator. Extraordinary, powerful communications systems have brought people of widely different basic disciplines together, merging their talents for the purpose of generating new and unique knowledge domains. What does the future hold for these pioneers, and what kind of leadership is required to manage these elements of powerful thinking?

Multidisciplinary programs have been around for a very long time. Quite frequently they have comprised segments of two or three basic disciplines. Examples of these would be biophysics, management science, agricultural science, bioengineering, computer engineering, engineering physics, and engineering management, to name a few. Most consist of loosely interconnected subsets of materials taken from each of the basic disciplines. Courseware or curriculum threads unique to the multidisciplinary studies often do not exist, at least not in significant quantity.

The notion of integrated studies surfaced when members of the business community were asked to identify the most important attributes of existing employees or new hires. The term "broadly-based" was near the top of virtually everyone's list. Other key indicators were "adaptability" and "flexibility," along with "the ability to learn quickly." Strong problem-formulation and problem-solving skills were on every list, and inevitably, the requirement that the individual must have a demonstrated ability to act as a team member or, even better, a team leader. Ironically, while these represented the major qualifications demanded by the employers and managers for their employees, organizational structures in the form of the time-honored organizational charts, position descriptions, and duties remained unchallenged. This was true not only for the private sector, but certainly for government agencies and educational institutions as well. This inconsistency was eventually recognized and actions in the form of re-engineering projects began to dominate the management scene.

In a certain sense, the marketplace was redefining the basic elements of job skills needed under new operational structures. Interesting examples came from the banking industry, where bankers noted that their business was being caught up in the new technologies of electronic commerce, encryption of data, and forward planning using neural networks, all in addition to the traditional financial skills and management requirements. Consulting firms showed an early interest in students with backgrounds in integrated studies, since their clients were often scattered among many different disciplines and business markets. Educators, of course, were quick to notice how these demands were being translated to curriculum reform for their undergraduate students. Educational institutions, like business, had to adjust to a new way of viewing their curriculum and even their organizational structure in the light of student courses and curriculum that crossed departmental and college boundaries.

While the move was on to develop a response to the demand for students with integrated studies experience and employees who were stronger team workers and more adept at accommodating new, more demanding job responsibilities, a concurrent evolution, or at least popularization, of "knowledge-based systems" was emerging. For example, expert systems brought recognition to knowledge-based systems as powerful tools that encourage higher level thinking, problem definition, and problem solving. Initially, the expert system concept was associated primarily with the building of a type of diagnostic machine, essentially an automated way of keeping track of a set of rules and data to steer a user through a diagnostic pattern. Expectations of artificial intelligence brought unjustified doubts for some time, but success stories built upon one another and the role of expert systems was established as a strong component of contemporary problem solving.

THE KNOWLEDGE AGE

For centuries, society has been driven by issues of geography and energy. These will continue to be powerful forcing elements. However, added to the mix will be the issues raised by the emergence of the "knowledge age." Knowledge is increasingly being considered as a commodity wherein ideas are formulated and distributed on a massive scale. The Internet has demonstrated the hunger for broad and open communications with access to all kinds and types of information. Educational and organizational systems are responding to the challenges and opportunities brought about by wide-sweeping electronic communications and the applications of knowledge-based systems.

The extraordinary changes introduced by a knowledge-driven society have been artfully articulated by Dr. George Bugliarello.[1] He views knowledge-based communities as evolving into power structures where geographical, economic, and political boundaries become transparent. This perspective brings even the definition of national sovereignty into discussion.

> In terms of generation of knowledge, or, in a narrow sense, of the extraction of information from natural or human-made events, the predominant organized players today are researchers and research institutions. The knowledge these players acquire is generally recorded and transmitted, albeit not universally received. Indirect measures of its magnitude show it is growing exponentially. In addition, there is a vast if poorly tapped body of valuable knowledge being generated outside the research laboratories, by experience, by trial and error, serendipitously. This largely grass roots knowledge is recorded far less systematically, if at all, is transmitted haphazardly, and is impossible to measure even indirectly.

In order to deal with this "transparency" of time and place, new educational structures must be designed and developed. Multidisciplinary or "integrated" studies will emerge primarily because of the ease with which elements of knowledge will be packaged and made accessible. Individual disciplines, of course, will persist, but interactions between and among the disciplines will take on new value. As the economist Boulding stated in 1968:[2]

> One wonders sometimes if science will not grind to a stop in an assemblage of walled-in hermits, each mumbling to himself words in a private language that only he can understand. In these days the arts may have beaten the sciences to this desert of mutual unintelligibility, but that may be merely because the swift intuitions of art reach the future faster than the plodding leg work of the scientist. The more science breaks into subgroups, the less communication is possible among the disciplines, however, the

greater chance there is that total growth of knowledge is being slowed down by the loss of relevant communications. The spread of specialized deafness means that someone who ought to know something that someone else knows isn't able to find out for lack of generalized ears.

The "generalized ears" referred to by Boulding may well be seen in the emergence of powerful knowledge-handling systems from the field of artificial intelligence. Expert systems development comes to mind because of its simplicity and its outstanding record of accomplishment in being the "generalized machine" that listens to information and knowledge and then proceeds to sort, structure, and reprocess it for the particular needs of the user.

Intellectual capital has previously been defined as the patents and trade secrets held within the corporate enterprise. More recently, the term includes all of those ideas, concepts, and rules that employees have accumulated over years of work and study. Thomas Stewart[3] summarized this notion in an article entitled "Brain Power." He writes:

> Most companies are filled with intelligence, but too much of it resides in the computer whiz who speaks a mile a minute in no known language, in the brash account manager who racks up great numbers but has alienated everyone, or in files moved to the basement. Or it's retired and gone fishing. The challenge is to capture, capitalize, and leverage this free floating brainpower. Every company depends increasingly on knowledge — patents, processes, management skills, technologies, information about customers and suppliers, and old-fashioned experience. Added together, this knowledge is intellectual capital.

Here we see, in the view of 3 individuals spanning 30 years, the same conclusions relative to the detriments of isolation and poor communication within an organization as contrasted with the opportunities that exist when knowledge is viewed as the most important commodity or core competency the institution possesses, provided, of course, that it is intelligently and effectively utilized. In the material that follows, we will outline how the emergence of knowledge-based systems has caused a revolution in the thinking of the private sector and education as well. We have chosen examples from expert systems primarily because they represent, in the form of hundreds of working examples, how complex multidisciplinary problems have been given simple, powerful structure and documentation. In this sense, we viewed the expert systems as a type of "thinking paradigm" for the development of what we have defined as "integrated studies." Organizational restructuring has been an especially popular subject for discussion in the private sector in recent years. Much of this effort has been the result of a critical need to define the core knowledge of the corporation and how it must be effectively employed to the competitive advantage. Old processes and outdated organizational charts have come under assault as being obstacles rather than positive mechanisms for handling the flow of knowledge within the organization. The notion of how knowledge is acquired within the corporation, and how it is assembled and restructured or repackaged for the user, often represents a competitive advantage for a company or an educational institution. Computers, through powerful inference engines and information handling capabilities, provide the essential tools for structuring and making available the institutional intelligence.

Applications of Total Quality Management (TQM), concurrent engineering, and similar pioneering efforts at optimizing decision-making, even in the elementary and simplest forms, illustrate the potential for knowledge-organizing systems.

KNOWLEDGE AS A COMMODITY

Society has recognized the value of knowledge for centuries, but certainly has not been very effective in determining methods for acquiring, classifying, measuring, distributing, and utilizing it. For example, 30 years ago new mathematics and free form curriculum were the avante garde techniques of educators. In the 1990s we are attempting to define national goals for our educational system, but are having difficulty reaching consensus as to what the performance standards of students should be and how they ought to be measured. We offer the Graduate Record Examination and the Student Achievement Tests as benchmarks of student performance, but admit that they do little to predict the students' potential for success in their future studies or careers. This is not meant to demean our present efforts in this area, but it does suggest that there is much we have to learn about how we view knowledge. The concept of many working in the field of knowledge-based systems is that knowledge can be viewed as a "commodity." This leads to thought patterns, in education at least, that are less skill-based and more oriented towards knowledge acquisition and knowledge generation in higher level thinking.

The National Assessment of Educational Progress (NAEP) reports that students' basic skills capabilities may be improving, but higher order thinking and independent learning are not. A project reported on by Scardamalia and Bereiter[4] describes a process used in a fifth-grade classroom that focuses on knowledge as something that a student recognizes as being acquired and shared. In this project, the computer is used to help create the learning environment where students can define and demonstrate what they have learned to their classmates. This, of course, represents a reinforcement of what all educators look for — student-to-student enhanced teaching. In this project, it seems clear that students and teachers treat knowledge as an object (or a commodity). This structure and environment looks much like that used by teams of professionals with distinct knowledge domains working together to create a problem definition or solution important to their organization. The authors of this study are quick to point out that their knowledge-based approach is quite a novelty to the students and to their teachers as well. As curriculum and courseware projects of this type gain in popularity and use, the powerful concepts and disciplines of knowledge-based systems will be seen as essential tools in bringing structure and compatibility to the educational system.

As experimental work in this area continues, one will see an increased awareness of the values of student-to-student teaching, constructs that enable personal and shared knowledge bases, and knowledge-base concepts that monitor the teaching/learning efforts as they move towards quantification of student performance criteria.

In higher education, the focus on knowledge concepts is seen in various programs developing courseware or entire curricula in "integrated studies." These projects are at least in part a response to the NAEP reports criticizing students' lack of independent learning. Curricula in integrated studies depend heavily upon computers and network systems in support of the courseware. In fact, much of this work looks as though it has been developed under the protocols established for the design of knowledge-based systems. The integration of computers into the curriculum clearly takes many forms, ranging from the development of reading and writing capabilities to the senior capstone work involving complicated research projects. The goal in most of this work is to move towards higher levels of student participation in the learning process and away from the pure skills or passive student behavior. The design of the curriculum requires a meticulous coordination of its components. These components may be special laboratories, reading assignments, independent computer-based instruction, writing, or verbal and multimedia communications. Participating faculty, of course, represent a richness of knowledge domains as teams of faculty begin the process of structuring integrated courseware. The curriculum that results is hypercarded, in effect. The student moves in and out of knowledge domains supported by the experts on the faculty. At the same time, other faculty expertly trained in knowledge-based systems orchestrate the continuity and the "integration" of these studies towards the goals and objectives established for the students. If this concept is to work well, the full power of knowledge-based systems disciplines will serve as the glue that holds the many components of this type of curriculum together.

JOBS, CAREERS, AND CURRICULUM DEVELOPMENT

The American Association of Governing Boards (AAGB), in its Winter 1996 edition of *Priorities*,[5] summarized the most important responsibilities for governing boards if they hope to make positive quality changes in higher education. Their conclusions, of course, apply equally well to all elements of education, including K–12, vocational, and post-degree professional development. In comparing our present conventional teaching and learning process with one proposed for the future, some strikingly new objectives for curriculum and courseware design take shape.

For the most part, our educational structures are dominated by disciplinary paradigms. In a quality-oriented process, according to AAGB, where consumer concerns are paramount, the curriculum would be driven by student needs and tested by feedback from students as well as employers, professional associations, and alumni. This observation reflects the weakness of disciplinary isolation and the strength of integrated studies. This, of course, comes as no surprise to those in the private sector who have been working on the integration of processes and people for the last decade or so. Usually under various discussions of re-engineering the organization, or the modeling of processes, one finds the ultimate objective to be that of preventing intra-organizational isolation while promoting cooperation and communication throughout the organization. The management styles and tools necessary to integrate the functions of the organization are growing in breadth and strength as technology develops. The point is, our students will enter this environment and be expected to understand the meaning and the significance of the words *multidisciplinary, integrated studies, concurrent processes,* and *knowledge-based systems.*

The success of integration has already made its mark in the private sector and most assuredly this thinking will impact our educational systems in a way we have not seen before. Faculty will see their independence challenged by an interdependence with their fellow faculty members. Teams of faculty will emerge, and in the process faculty will find their role as educational process managers becoming as important as their background as a content expert was a few years ago. The curriculum, the structure, and the course content will undergo a dramatic change, primarily because the delivery process required in educational systems of the future will be as important as course content. This change again reinforces the need to be thinking broadly in the development of an integrated educational process. The notion of "just in time" and "just enough" will become a powerful problem-solving and critical thinking strategy.

We have identified above the following important goals for education:

1. The curricula must be driven by student needs.
2. Courseware will show an integration of disciplines with reinforcing elements.
3. The delivery process will be given equal consideration to course content.
4. The emerging educational systems will be managed by faculty members and teachers working in a team environment.

Each of these goals requires the innovative and effective use of technology. This vision removes the perception of computers on campus being synonymous with computer laboratories and scattered work stations. Instead, we have created a vision that demands the integration of curriculum, networks, computers, students, teachers, and technicians as the essence of "computation on campus."

The development of the curriculum to support the kind of integrated study discussed above is a task unfamiliar to the majority of teachers and faculty. Contemporary course work in environmental studies, biological sciences, telecommunications, multimedia communications, computer-based instructional programming, public policy, and ethics all require participation by experts in the constituent fields. As courseware is developed by these teams of faculty, each member is in an intense learning mode. Each member of the team must determine the nature and level of his or her contribution to the overall objectives of the courseware. This is a process of selecting the **strategic** elements of knowledge required for the task at hand and making this knowledge and information accessible and available as appropriate to the requirements and needs of fellow faculty and the student. Knowledge-based systems play an essential role in this integrative process in that they allow the teacher and the learner to reach common objectives about the degree and extent of the content material required. Access to knowledge bases, both local and remote, allow the students to select where and when they wish to do in-depth studies in an efficient and timely fashion. There are certain fields where the computer is an essential, powerful enabling tool. The modeling of complex systems is one example. Exploring computer-generated fractals is another opportunity now available to the student particularly motivated towards further studies in visualization, art, and quantitative methods.

Another important development as curricula progress into integrated studies is the development of stronger course work in communications. Written and verbal communications must still be mastered; however, the elements of presentation strategies and the visualization of complex situations and data in a multimedia environment are becoming essential to virtually any profession.

Finally, those curricula that require a senior year capstone project are demonstrating that they have an effective assessment vehicle for their program. In addition, students that have a recruitment portfolio with a strong project orientation are often able to discuss and demonstrate their awareness of the importance of team building, searching for information and knowledge domains, and mastering effective communications and presentation strategies.

Very practical and productive work is being done by faculty teams in higher education. Early in May of 1996, three institutions participated in a workshop designed to encourage discussion of how faculty were organizing and offering curricula in integrated studies. Faculty from James Madison University's College of Integrated Science and Technology discussed the development of their curriculum. This curriculum is in its 3rd year with students now completing their junior year. Approximately 50 students started the program in 1992. The college expects the number of majors enrolled in the program in the fall of 1996 to be close to 500. The number of faculty has risen from essentially 0 to more than 30 in 4 years. The experiences, both administrative and academic, of initiating this kind of new program received considerable discussion, demonstrating how the roles of the participating faculty in this program differ greatly from those in single-discipline departments.

At Marshall University in West Virginia, the planning process for a new bachelor's degree program in integrated science began in 1993. Faculty teams were organized to develop the structure of the curriculum and define the long term development plans for the program. This program is similar to that at James Madison except that it has a stronger element of management, more active participation from the liberal arts and fine arts, and has fewer options for senior studies, concentrating mainly on environmental studies, manufacturing, and telecommunications. The program was approved by the Board of Trustees in early June of 1996. It is expected that the first class of students will enroll for the program in the fall of 1996.

Faculty from the University of Oklahoma at Norman were present to observe and discuss the possibility for a general studies program. It appears at this point that the direction of this curriculum and courseware may move to a program with less concentration on the hard sciences and an increased focus on the social sciences.

As work progresses in the development of new courseware and curricula, several factors must be given serious study. First, we must continually ask what kind of form and configuration computer systems must take in order to present accessibility and functionality to the educators and students. Cost will continue to be a primary factor in this consideration; therefore, those programs that support large and complex computer laboratories must realize that this investment is tentative because of the nature of the change in technology and because of the large overhead in personnel and software that is needed to support the laboratories. Large computer laboratories

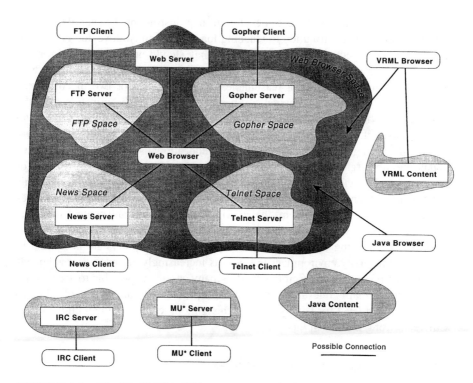

FIGURE 1.1. The World Wide Web represents a classic example of how hardware and people are integrated. Text (Gopher) servers and file transfer (FTP) servers are merged with virtual reality modeling language (VRML) and, multiuser (MU) systems, the Internet Relay Chat (IRC) servers, JAVA, and news clients. This represents the working reality of any time, any place, any pace knowledge acquisition. (From TeleGeography, Inc., Washington, D.C. With permission.)

with weak links to curriculum requirements constitute a nonproductive overhead and cannot be tolerated in times where money is short and efficiency is demanded. Better management of the technological resources on campus is required, and too often this falls to informal, powerless committees.

The ideal sought by some institutions is to require students to have their own portable computers. Institutions would then be primarily responsible for providing the local area networks, high level software licenses, and student docking stations or wireless connections to local servers. The discussion concerning "appliance"-type user access must be seriously considered. Many of the very powerful Internet tools make searching, browsing, and authoring very attractive. The advantages of multi-platform communications, as well as a common software base for users, are also clear advantages of centralized software. One would expect intranets to evolve rapidly in the academic environment.[6,7] (Figure 1.1)

While the discussion of the applications of high technology in education continues to be broad and complex, it is clear that the development of electronic-age instructional materials is not keeping pace with the demands that society is placing on education at all levels. Faculty developing new instructional materials will not be able to make the contribution to teaching loads or administrative and committee responsibilities expected of other faculty and they should not be penalized for this difference. Further, the approval process for course and curriculum change needs to be vastly streamlined. It is not reasonable to expect a year or two of committee discussions for each course change. As has been shown in the private sector, the most agile of the organizations will be offered extended opportunities for growth and development.

We are moving toward the development of educational systems that will accommodate the time schedules, the content, the interests, and the career directions of the learners. The emergence of distance learning and the major efforts in this direction is only one indicator. Everything from e-mail to the Internet characterizes the learner independent of systems that are rigid in time, place, and content structure. The interdisciplinary traffic that is commonplace in the job market is not the traditional way of working on many university campuses. Computers are beginning to change the university environment in that it is increasingly easier to access knowledge outside of one's own discipline, and in many cases it is possible to access only that knowledge that is relevant to the particular problem at hand. This is leading to a higher comfort level for faculty that are engaging in team teaching within courses with very strong interdisciplinary content.

THE EVOLUTION OF EXPERT SYSTEMS

Initially, expert systems were developed on the basis of need by those trying to troubleshoot defective equipment or perform a medical diagnosis from a set of relevant symptoms. However, as applications grew in sophistication, they also spread across many professional and disciplinary boundaries. The process of developing expert systems forced a mixture of the talents of people taken from many knowledge domains. Problem formulation and solution was the objective of people working as a team sharing their knowledge. Structure, control, and coordination were brought to the **process** of knowledge storing and building. Bringing problem context and strategic information relevant to the problem at hand demonstrated a powerful technique for solving problems. Equally important, the priceless objective of achieving concurrence between team participants developed in an amazingly short period of time, often in an environment that promoted trust and respect.

In earlier years, many of us participating in such exercises were concentrating on using the powerful new tools and skill sets of knowledge-based systems to solve critical, complex, yet often simple problems. As additional experience was gained, it became clear that the entire arena of knowledge-based systems, including expert systems development, presented a new and refreshing way of applying an eminently powerful process to the issues of team-building, interdisciplinary problem-solving, concurrent engineering, complex project management, and TQM, to name a few. This common environment of people and problems that the expert system addresses demonstrates continually and graphically how knowledge can be shared and how new knowledge can be developed, all in the context of "just in time and just enough" strategic knowledge to accomplish the task at hand.

The expert system experiences also demonstrate the intellectual power that is unleashed when one looks at problem formulation and solution independent from a preconceived and often incorrect notion of the mathematics that must be brought to the problem. With the expert systems, one often takes the inference engines for granted and focuses on the modeling expertise and simulation concepts that lead quite naturally to a mixture of deterministic mathematics, with its equations plus a vast array of table look-up and database information, all interactive with rules generated from a knowledge gained from individual intuition and experience. For many of us in the system sciences, this presents new opportunities for solving practical problems of increasing breadth. Equally important, participants in these problem-solving sessions begin to move beyond the paradigms that have been developed by a strongly structured educational and professional environment. This is a difficult transition and one that not all team members can finally accommodate. Indeed, this cannot be considered a liability for either the team or the strongly discipline-oriented individual. It simply indicates that there is more than adequate opportunity for future development for both the specialist and the generalist. What team-building with knowledge-based systems demonstrates is that powerful new tools have been made available by work in computation, particularly artificial intelligence, that offer to the generalist unique abilities in the marketplace as we move rapidly into the age that many have identified as information-intense and discipline-transparent.

The contents of this book illustrate the broad boundaries of knowledge-based systems. For some, it represents the opportunity to concentrate on the further development of inference engines. For others, it offers opportunities to explore the rapidly expanding field of knowledge elicitation. For some, it brings the power of computation solidly into the hands of managers, while for others it offers the power of multimedia communications and interactive learning directly to the users of the end products developed from knowledge-based systems applications.

Topics such as re-engineering, reorganization, career and job changes in the work force, and ever-increasing expectations for computer literacy are commonplace in business, government, and education. Employers continually search for the "team player" with "excellent communication skills." It is unstated, but certainly implied, that these two skill sets should be combined with strong abilities to learn, to teach, and to excel in one or more knowledge domains. The first two of these requirements are seen to be strong components in what has been defined as "electronic communities." The term refers to those individuals with common interests who have come together largely through electronic means to share in the acquisition and distribution of information. If one doubts that such communities exist, look to the Internet for convincing arguments. Essentially, the Internet is an unstructured system where individuals browse in search of a community of participants that share their particular interests. Expert systems projects often created their communities of participants because of a common problem-solving goal. The participants came from widely separated geographical and disciplinary sources, but were brought into a community of teachers and learners all dependent on each other.

There are thousands of instances where artificial intelligence, and specifically expert systems, has made long-term, dramatic change within an institution or organization. Many are not highly publicized, some for the obvious reason that the commercial value or competitive advantage brought about by the application was too good to share.

What follows are three examples of expert systems that dramatically show the value of expert systems projects in managing "communities," integrated studies, and knowledge constructs. They are simple in concept, relatively easy to implement, and have caused significant change in the way business is conducted. These examples are offered to illustrate the sometimes subtle, but very powerful, process of thinking and working that generates out of the experience of building an expert system. In each case, the elements of knowledge elicitation, followed by the restructuring and repackaging of knowledge, and finally by the presentation of knowledge, is a common architecture.

An application that dates back to the early 1980s involved a simple configuration monitor for the sales, production, and delivery of computers, networks, and printers being delivered by Telex Corporation. As sales dramatically increased in the company, there emerged the recurring problem of certain products being delivered and identified as unacceptable by the customer. Almost all of the reasons were extremely simple matters of mishandling and had very little to do with product performance. Mistakes included delivering incorrect operational manuals, supplying improper interconnecting cables, sending a network card with minor configuration problems, and delivering the wrong product at the wrong time. An expert system was built to serve as the ultimate master control template between the initial execution of the sales order and the final checkpoint of product configuration. Simple prompts were given through the process to make sure that past errors were being corrected and that potential for error or oversight was being highlighted. The application became increasingly important as customers began to order widely diversified machine and network configurations. Past procedures relied on manuals for sales, production, and delivery that were often changed and, therefore, were loaded with footnotes regarding configuration exceptions for particular customers or classes of customers. This is a classic example of how a problem was identified, a boundary was placed on the context level of the problem, and information and control were established. The stage of eliciting knowledge was one of broad and open generation of ideas and concepts as to how the system should be run and what problems had to be avoided. This moved into the coding of the program that served to unify databases, place the appropriate rules into the structure, and design the interfaces to communicate the right knowledge to the right person at the right time to guarantee proper action. Like other such applications, this turned out to be a large money-saver, and its use in a modified form continues to this day. Obviously, many applications of this type have been developed, but too little attention is given to analyzing the nature of communication, team-building, and problem formulation that is offered by following expert systems disciplines.

In a second example, an expert system was developed for Facet Corporation, a manufacturer and supplier of filtration equipment. For years, the company had supplied a sensor and transducer component for the Department of Defense. The product had been continually improved and re-engineered over more than two decades. The engineering effort was conducted by three senior engineers over a large part of this time. Then management realized that these engineers would be retiring within 2 years. These senior engineers carried a great deal of the engineering design process, including minute details, in the recesses of their minds. Most engineering documentation is thorough, but not all details are recorded. After 6 months, it was concluded that a process would have to be developed which would provide a teaching-learning experience between newly hired engineers and the senior engineers. The process of developing an expert system was set in place, knowledge elicitation began, and immediately the "community" of participants defined for this project grew in size and depth. Vendors that participated in the design over the years were brought into the discussions, along with metallurgists that worked on the project, other engineers in the company that had contributed their work to the effort over the years, and, of course, the government customer became an active part of the process. The expert system was developed and delivered, and it met the goal of capturing the corporate experience that was needed to continue this product line. The point is less the fact that the expert system evolved than that this constituted a classic example of interdisciplinary studies, concurrent engineering, "idea field" management, and teaching and learning methodologies all packaged into one effort targeted at preserving corporate intellectual capital.

The third and final example is an engineering design involving the manufacture of an oil filter produced by Facet Corporation. Engineers had been designing filters for years using the concepts of fluid flow, filtration, and fabrication, to mention a few. However, as customers became increasingly demanding for products with new and improved specifications, greater quality levels, and lower costs, the re-engineering of the products increased in complexity while engineering time-to-deliver was profoundly decreased.

An expert system was put in place that provided a very simple behavior model for the design engineer, enabling a dramatic decrease in engineering time for prototypes. The system provided a highly interactive graphical interface to the engineer, allowing for the complete interactivity of performance functions for the filter together with cost and manufacturing constraints. The system was developed in the mid-1980s and is still in use today. This system, like many expert systems, involved the input from a team of individuals having widely diversified responsibilities in the company, not only engineers, but suppliers of components, customers, individuals from accounting and costing, sales representatives, and engine designers, to name a few. These participants contributed to the project by providing their expert knowledge in the form of the appropriate strategic content needed to fit the requirements of the problem at hand. Individuals crossed the disciplines of finance, engineering, and sales, each learning the other's jargon and, most importantly, appreciating the significance of this project to all of those participating as well as to the company itself.

These are three simple projects that had a very significant impact, not only from an economic point of view within the company, but more importantly, as a mechanism and process for controlling how individuals with widely diverse backgrounds and responsibilities can be brought together into a well-formed team process. There probably have been failures in attempts at building expert systems, but more than likely these are due to overly ambitious initial expectations, a lack of continuing commitment, or most probably, from a static and intransigent management structure responsible for implementation. There remain managers that have not moved beyond the black microphone, the manila file, and the sanctity of the organizational chart. In these instances, the threat of loss of control is too much to overcome. However, when companies assert that they are looking for new employees with strong communication skills and expanded computer capabilities, and who have demonstrated that they can work as a member of a team, it is clear that they are defining the kind of an employee that already understands and has been exposed to the elements of knowledge-based systems. There are too many managers that see expert systems as simply a device to create some type of diagnostic or evaluation program. It is far more than that. The very process of developing such systems teaches both the art and the science of problem formulation, the methodologies of research, the necessity of discipline, dependency on others, and the power of persuasion and communication. Strangely, corporations that have benefited greatly from knowledge-based systems applications seem to have failed to recognize the profound intellectual depth offered up by this process. High level management follow-up is too often feeble at best.

AN INTEGRATED SCIENCE AND TECHNOLOGY CURRICULUM

In 1988, James Madison University began a study of how its degree offerings might be extended into areas of the applied sciences, engineering, and technology — a mission and role change that many predominantly liberal arts institutions might face in the future. The decision was made to create a College of Integrated Science and Technology offering a bachelor's degree by the same name. A blue ribbon committee was assembled to outline the formal structure and purpose of the college.[8] Dr. John H. "Jack" Gibbons, currently the head of the Office of Science and Technology Policy and advisor for science and technology to President Clinton, was co-chairperson for that committee. Since those beginnings, land for a new campus has been purchased, a master campus plan has been completed, and the architectural designs and specifications for the first three academic buildings have been finished. The first building will be occupied in 1997. The programs in the college initially included degree studies in integrated science, computer science, and health and human services — all professional programs with a great deal of common interest and structure from the technical to the delivery of services.

Faculty and corporate advisors looked carefully at the changing job and career market and identified areas not being served by the university. As the curriculum developed, it was clear that governmental and private sector organizations of a very broad range were interested in participating in this project. Interest came from the government, banks, engineering and consulting firms, small businesses, manufacturing organizations, environmental and public policy organizations, high tech industries in computer science and information systems, and from telecommunications operators. The objective was to develop a core curriculum with options that would serve the needs of such organizations. The new Bachelor of Science degree in

Integrated Science and Technology represents a response to an important role that will be played by individuals who can bring the community of science, technology, and engineering into closer understanding of the economic, social, and political issues that desperately need technical explication.

While the curriculum was always intended to have a strong technological and scientific orientation, it was also realized that a number of students, particularly women and minorities, were not being attracted to the traditional science and engineering disciplines and specializations. The freshman and sophomore years were structured to offer courses that integrated many of the fields of science with engineering, computer science, and knowledge-based systems studies. The junior year presents a core curriculum intended as further study of the integration of the strategic areas in science, biotechnology, instrumentation fundamentals, knowledge-based systems, manufacturing, engineering, and management. The senior year focuses on one of six options for specialization.

Students graduating from this program will develop knowledge and skill in three principal areas:

1. Problem-solving
2. Communication
3. Sensitivity to context

The curriculum design serves to:

- invert the learning progression of traditional science and technology by moving context and applications to early courses,
- integrate issues of global commerce, government studies, and business through instructional modules developed by faculty of many disciplines,
- define a new degree experience that integrates the areas of science, engineering, computer science, knowledge-based studies, management, analytical methods, and liberal studies, and
- identify the importance of science and technology in the context of societal needs and issues throughout the 4 years of undergraduate study.

The structure of a similar program developed at Marshall University in Huntington, West Virginia, is shown on Figure 1.2. The heavy technical concentration at the freshman and sophomore levels is aimed at engaging students, particularly women and minorities, in the discussion and discovery phases of problem definition and problem solving, with science and technology in context. The core sector defines studies that serve as the framework under which broad concepts of science, engineering, management, economics, and social factors interact. The challenge is quite obviously one of developing the structure and the process through which this content will be articulated to the student: 18 hours of specialization are indicated, 6 hours of which are devoted to the senior thesis. Also, 12 hours are provided for extended studies in one of the sector areas or for further inquiries into areas of related interest. Course content of these areas includes requirements in writing and presentation skills, use of knowledge-based systems, and the integration of business concepts.

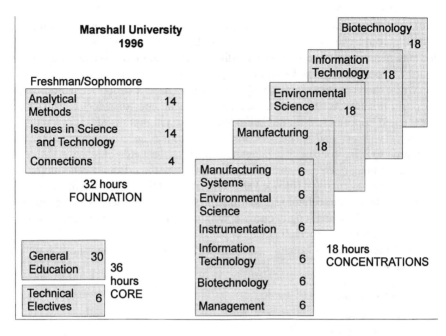

FIGURE 1.2. This diagram illustrates Marshall University's newly approved Integrated Science and Technology bachelor's degree curriculum. Note the 32 hours of foundation courses are strongly integrated studies in science, computation, modeling, and mathematics. Majors, or concentrations, are initially aimed at manufacturing and the environmental sciences.

The quality indices for the curriculum are based upon: engagement of students with science, integration of disciplines, use of advanced communications concepts and context-based study, focusing on problem solving and critical thinking. This is a very different approach to the study of science and technology in that it is broad and multidisciplinary. Further, by being presented with science in the context of real world issues, students are able to see the significance of technology within the complex and sometimes disorganized environment of contemporary societal and economic issues. Typical of knowledge-based systems good practice, courses employ expertise from the field of modern communications where concepts and ideas are reformatted for most effective communication between machines and people.

This curriculum development at James Madison University started in 1992 and refinements continue. To ensure consistency in the development process towards a cohesive curriculum, specific objectives were established:

- Initiate a new crafting of the integration of science, management, liberal arts, knowledge-based systems, and analytical methods.
- Create a more effective method of engaging people by presenting the value and excitement of science and technology.
- Define a new role for faculty with the ability to explore new teaching methods using advanced communication vehicles across disciplines.

- Create a sensitivity to cost containment concepts by using previously developed instructional materials, gaining access to external laboratories, hands-on experience, and expanded use of simulation.
- Prepare a greatly expanded consideration of advanced communications (machine to machine, man to machine, person to person) all brought to this program through the involvement of communications faculty contributing to the design and development of science and technology curriculum.
- Emphasize teaching approaches particularly sensitive to the positive and productive ways in which minorities and women can be encouraged to identify and seek careers in science and technology.
- Renew emphasis on how courses can offer a broader and more effective educational process in helping students to identify and select career paths.

Liberal Studies are offered in all 4 years and may include courses such as Technology and Western Society, History of Modern Science, Early Modern Europe: The New Worlds of Exploration and Science, Problems in Medical Ethics, Philosophy and Scientific Inquiry, Technology and Sociology, and the Economic History of the United States. Courses also include Fine Arts and Aesthetics, Literature, Oral Communication, and Composition.

Strategic sciences are presented through an **Issues in Science** component that is offered in the freshman and sophomore years. This component is designed to **engage** the student with science by offering an integration of the sciences presented in the context of realistic problem-solving. In addition, by being provided a real-world project context, students are exposed to important levels of **career choice** options.

The third component of the curriculum is **Analytical Methods.** A freshman and sophomore track, this component is designed to teach the ability to analyze an issue, postulate a model, formulate a hypothesis about the problem, decide what variables to measure and how, design an experiment or survey, collect data, test sensitivity of the model, interpret data, and draw conclusions. Students will learn analytical methods in studying fields like visualizing information, the calculus of science, research methodologies, reasoning with expert systems, and design of experiments.

The fourth component of the curriculum is the **Science and Technology** sequence. This is part of the junior and senior year, and is intended to engage students in the dynamics of interdisciplinary activity by putting them in context-driven "sectors." Examples of initial sectors offered are as follows: engineering, management concepts, manufacturing, instrumentation, knowledge-based systems, computers and computation, and information science.

This degree program provides extensive interdisciplinary exposure and training that (1) provides a strong new initiative for recruiting and preparing students for careers in science, technology, and engineering-related fields; (2) prepares students for the "real world" of complex multidisciplinary issues; and (3) encourages a stronger engagement of science and technology with the public in general.

The success of programs of this nature depends upon the application of a cohesive influence continually working to integrate students and faculty within a context where knowledge is treated as an object or commodity.

Thus, knowledge-based systems will serve as the integrative mechanism. Knowledge-based systems provide the framework for handling the exchange and integration of knowledge from various sources. It forces knowledge bases to be created for ultimate sharing and analysis. Knowledge-based systems are an ideal technology for capturing, preserving, and documenting knowledge. Knowledge-based systems can be very useful for logging the teaching-learning experiences so that continuous improvement will occur and the "intellectual capital" of the effort will be systematically preserved.

Computing systems are changing the university environment making it easier to access knowledge outside of one's own discipline, selecting knowledge that is relevant to the particular problem at hand. This is leading to a higher comfort level for faculty that are engaging in team teaching within courses with very strong interdisciplinary content.

As technologies deliver faster and better ways of assembling instructional programs, faculty will respond by developing systems of instructional modules that can be used by a variety of students participating in a networked knowledge-based system. We know what must be done. As we have stated, tools are evolving rapidly to accomplish this "field of ideas" environment. But we are still learning how painful it is to leave one's own paradigm of the educational process and be greeted by a host of new, complex, and threatening concepts that serve the purpose of collecting information, creating knowledge units, eliciting knowledge, repackaging knowledge, and finally delivering it in the most effective method possible.

In the short term, programs of this type will gain popularity because of the market strength for graduates of this kind of curriculum. However, the long term strength and stability will be measured by the extent to which faculty recognize and utilize the powerful integrating process of knowledge-based systems.

COMMUNITY

For the longest time we have known that communities of people sharing common interests, whether of formal or informal construct, make incredible contributions to society. In the past, such communities were largely based on geographical boundaries. Communications and computers have drastically changed the way communities of common interests are evolving. In particular, the Internet, with its free-form constructs and open, virtually rule-less, society, has spawned the growth of very productive interactions between individuals, corporations, and the government. Notable is the present dramatic move towards the increasing commercialization of the "Net." The future will show us an entirely different form of commerce (electronic commerce) than we have ever imagined, much less planned for.[9] Communities, in a unique way, demonstrate the unconstrained and unbounded power individuals and/or organizations play in an "idea field."

The organizational chart has been, and continues to be, in many cases, the communication and idea network in organizations. The people and their ideas are constrained by a rigid organizational structure, one that too often is static and outrageously bureaucratically controlled. The Net, on the other hand, is at the other end of the spectrum, with very few rules and with an open door to communications and ideas from virtually anyone, any time. Clearly, progressive organizations that have been questioning their allegiance to the organizational chart are turning to the concepts revealed in the Net to look for new options. A product of this search is, of course, the "Intranet." Here the concept is one of promoting active and productive vertical and lateral communication within organizations, but under the control of operations and structure, including security precautions, that provide for profit.

The recent attention being paid to the notion of "community" is nourished by the activities of the Internet. In point of fact, the Internet was first developed as a communications structure for multidisciplinary teams of scientists, but as communications technologies advanced, so did the information handling power of many users. Therefore, over a period of 20 years, we have seen the concurrent development of (1) common interest communities, (2) wide and powerful electronic communications, (3) powerful user access machines, and (4) powerful concepts in programs from artificial intelligence research that manage knowledge as a commodity. The notion of communities has to be taken as a very pervasive element in this mix of four partners, for it is here that purpose and focus take form. In education, we see communities constantly developing among faculty members crossing disciplinary boundaries to work together. There are also communities of educators dedicated to the continual evolution of the best of classroom instruction. There will be others that make their commitment to the delivery of education off campus and outside the classroom. Still others will find themselves playing a stronger role in the counseling and coaching of students in an increasingly complex pattern of educational opportunities (Figure 1.3). Institutions that have developed new courses and degree programs in integrated studies are already seeing new communities of common interest developing among faculty members as well as between faculty and students (Integrated Science Degree Programs, James Madison University, Harrisonburg, VA, and Marshall University, Huntington, WV).

Watching the present growth of on-line communities as the private sector moves deeper into the world of electronic commerce will give indications of how we will be forced to deal with the information-based society. We will be continually faced with the search for and definition of community development, and this work may well be led by those now designing and packaging intranet configurations. Armstrong and Hagel[10] provide some interesting insight into how one could manage the development and evolution of the on-line community. First, if there are goods and services to be provided, one needs the business structure to handle this activity. Secondly, attention must be paid to the continual interactive feedback that occurs among the membership. This must be encouraged and focused, a role for a facilitating leader in the community. Further, attention must be paid to the overall look and feel of the community, led by someone akin to an editor or publisher that works to make the user interface productive and encouraging to the members. And finally, a new expert must be in charge of the routine operations, such as maintaining reference holdings, planning for the future, and collecting and repackaging information and knowledge concepts.

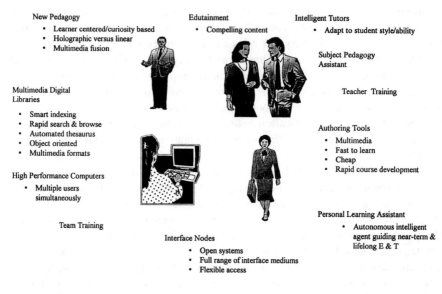

New Pedagogy
- Learner centered/curiosity based
- Holographic versus linear
- Multimedia fusion

Edutainment
- Compelling content

Intelligent Tutors
- Adapt to student style/ability

Subject Pedagogy
Assistant

Multimedia Digital
Libraries
- Smart indexing
- Rapid search & browse
- Automated thesaurus
- Object oriented
- Multimedia formats

Teacher Training

Authoring Tools
- Multimedia
- Fast to learn
- Cheap
- Rapid course development

High Performance Computers
- Multiple users
simultaneously

Team Training

Interface Nodes
- Open systems
- Full range of interface mediums
- Flexible access

Personal Learning Assistant
- Autonomous intelligent
agent guiding near-term &
lifelong E & T

FIGURE 1.3. This diagram illustrates the breadth and complexity of the learning environment for students. It further reflects the changes that need to occur in academic curricula and courseware in order to take full advantage of the technical advances in how materials are prepared and, especially, how they are delivered.

These notions of on-line community development will be defined, structured, and implemented within the current context of the Net. However, new ideas and concepts here are likely to be powerful elements of change for a world that is currently in communications disorder. The specter of information overload shows that present techniques are generating methodologies for access to data far more quickly than we are able to evaluate it, restructure it, and repackage it for the people or communities that need it. In short, getting on the information highway has little purpose if we are traveling without cargo or points of destination. It is in communications that knowledge-based systems and computers will find opportunities to better manage our economic and social issues and provide stronger and more flexible educational systems with far greater access.

ORGANIZATIONAL DESIGN

In recent years, some heavyweights in scientific thought have created an arena of discussion centered on complexity, chaos, expert systems, and knowledge management foundations.[11,12] These works cause us to reexamine how systems (social, economic, technical, educational) behave. We tend to worship the organizational chart and brag about how precisely we have defined the work of every individual in the organization. This is not surprising since it reflects how we seem to view our man-made world as a collection of individual parts and pieces that, when assembled, form the machine identified as the organization (the business, the university, the bureaucracy, the laboratory, etc.) — the Newtonian model. Our comfortable model of the Newtonian methodology focuses on things rather than relationships. Therefore we build large, rigid bureaucratic structures that shield us from uncertainty. Natural systems are characterized by continual change and adaptation. Stability and predictability are found less often than uncertainty and change. Following the natural model we can view organizations as reservoirs of ideas and concepts that constitute the intellectual capital of the whole. Organizations have begun to capture, capitalize, and leverage these "idea fields" using knowledge-based systems as a major tool in meeting this need. These new concepts are being driven by relentless advancement in the technology surrounding computation and its associated technologies.

Organizations rich with networks that traffic ideas, new concepts, and improved processes possess the agility and flexibility to adjust to the dramatic changes that are present in both our social and economic structures in the 1990s. For those of us that believe that virtual companies and virtual universities are near-term realities, it seems obvious that organizational structures of the "fields" variety rather than the "machine" variety will be the rule rather than the exception.

Certainly, the 1990s have seen unprecedented attention to both the study and implementation of new organizational structures and institutional processes. Under such terms as re-engineering, downsizing, and right-sizing, we find approaches that amount to everything from "squeezing down" to complete disassembly and resynthesizing of how complete organizations or businesses operate. The most timid of these efforts have produced positive results even in cases where the product of the work was simply that of making existing processes better-defined, eliminating unnecessary and inefficient work, redefining jobs, and providing workers with more information to meet goals of improved efficiency. In bolder efforts, massive change has produced new, innovative, and creative management structures. In this process, **knowledge-based systems** are supplying the tools of intelligence that define how organizations are structured, how they respond and react to opportunities, and how vertical and horizontal communications encourage productive competition for resources while creating highly efficient orders of parallel processing (concurrent synchronized management) within the organization.

Entangled in all of the work in organizational restructuring is the persistent emergence of the term "integrated." When knowledge, information, and data were scarce or difficult to access, there were good reasons for defining systems that created narrow, rigid, disciplinary structures. This is seen clearly in the disciplinary structures that have evolved within higher education. Many of these were produced because of the convenience of clustering information and knowledge while grouping people with compatible capabilities and skills among the resources they need to delve continually deeper into their studies. Such targeted in-depth exploration obviously must and will continue. However, the aggressive work being done in the management of knowledge has made it possible for individuals to move out of their primary disciplines and into other fields where they have a welcomed latitude to select the information from those fields that is most strategic or most important to the studies before them. Recently, we have begun to see very powerful techniques evolving for knowledge elicitation, the repackaging of knowledge, and finally, the presentation (communication) of concepts and ideas through entirely new multimedia systems. These presentation schemes offer up a new means for creating teaching and learning experiences where context and visualization offer a much broader range of learning paradigms than have been developed and practiced in the past.

It should not be lost to those of us that have been involved in "re-engineering" efforts that knowledge-based concepts are fundamental tools for creating new organizational structures and processes. Extremely important is the continual development of a learning process for the managers charged with the responsibility for actually implementing the new environment of growth and change, including new information which is not readily accepted. Managers may not have the experience or "representational scheme" into which *knowledge can be assimilated.*

Particularly successful management styles encourage the building of flexible structures that allow individuals to participate in projects when they are needed, and only to the extent that they can make positive contributions. Such was the thinking of the TQM movement over a decade ago. However, many TQM efforts did not provide the foundation needed for long term success. Committees were established, focusing on customer service with ever increasing quality standards. Unfortunately, committee memberships changed, the meanings and standards of quality changed, and, of course, technology in every form and format changed. In effect, many of the failed TQM efforts were not structured to employ the "idea field" concept; rather, they were a static appendage to the traditional organizational chart mentality.

Managers who have not been exposed to the knowledge-handling concepts of knowledge-based systems will be at a distinct disadvantage as they attempt to consolidate and communicate the elements of the horizontal and vertical idea fields of the organization.

Some organizations pride themselves on their strategic planning exercises. Attention and discussion are given to this subject perhaps once a year. The meetings are classic examples of the "retreat" syndrome. A speech is given by key executives followed by the formation of committees or discussion groups assigned to focus attention on an issue of major significance to the company or institution. In the discussion groups, a recording secretary is appointed, usually someone who has temporarily stepped out of the room, and the flip charts and magic markers are dragged out. For the more advanced sessions, a laptop computer and Post-It™ Notes are used to facilitate a fluid discussion of the subject. The dynamics of the sessions are predictable. Usually not more than 20% of those in attendance are interested in making a contribution, and of those, only half do so.

About 10 minutes before the session is due to end, a summary is committed to the flip chart which will subsequently be moved into a larger room where it will join its companion flip charts from the participating work groups. A designated speaker from each group will present a summary while the recording secretary attempts to make a cogent and logical summary of the summaries.

The problem with this format is that it invariably takes an outside observer or facilitator to manage or even control the human dynamics of open sessions of this kind. Facilitators have the best job in this environment, since they accept the job only under conditions of no personal accountability.

The difficulty with the process described is that ideas are rarely dependent on time and place. Ideas are spontaneous and created out of the assemblage of convergent factors generating a new thought or concept. To be effective, the notion of organizational "idea fields" needs to be an ongoing process, and to accommodate this rather strong requirement, new enabling tools and skills are needed. It is here that we again observe the value of understanding the elements of knowledge-based systems. If we use expert systems development as an example, we see that the process of knowledge elicitation leads to the definition of our knowledge domain experts within the organization. As particular issues for study are raised, the appropriate knowledge domain personnel are assembled. From here, the problem formulation continues, where common databases are identified, operational rules and policies are defined, and the equations of constraints and possibilities are formulated. All of this moves to the expert system shell, where it is subjected to powerful inference engines for interpretation. The concepts and ideas feeding this system can come from anywhere in the organization at any time, since they are all structured and linked through the computers and networks hosting the integrative software and inference engines. As the process evolves, participants see the viewpoints of other users brought into the context of the issue at hand. Constraints are made quantifiable, and in this context of knowledge sharing, the sophistication and generation of new knowledge continues.

Such systems are relatively immune to the overly aggressive, authoritative, or pompous demeanor that can be presented in face-to-face environments. This whole concept represents "the idea fields" in action with the enabling mechanism being handled by the knowledge-based system serving as the facilitator.

Many organizations have been given national attention for organizational restructuring that involves a continuous and systematic reassignment of individuals to meet the most critical issues facing the organization. While this process represents the needed fluidity for the idea fields, it is not necessarily a well-structured and documented process. It is a move in the right direction, where flexibility replaces the search for optimum stability. This allows people, job functions, and processes to change and the generation of a certain self-optimization of the organization begins. The emergence of the idea fields is clear as important knowledge runs rampant across the various structures of the organization. As this develops, one can quickly see how the important components of knowledge-based systems are utilized, specifically knowledge elicitation, knowledge classification, repackaging, and communications.

PROFESSIONAL TRANSITIONING

The events of 1993 and 1994 have drawn consistent attention to issues of organizational restructuring in the private sector and reorientation of the mission and goals of many of our national laboratories.[13] In addition, the nature of the national defense structure is being redefined and, with it, career opportunities are changing. The security of managing one or two career moves in a lifetime has been replaced by an expectation of a job change every five years. Extrapolating this trend into the future defines a need for programs that produce students capable of foreseeing and adjusting to professional redirection. Enhanced self-study habits and multidisciplinary exposure are suggested. Professionals of the future will either demonstrate leadership potential to their organization by communicating their ideas effectively and forcibly or they will fall to the role of an expendable service function. A more comprehensive understanding of the world in which one works, whether it be government, education, or the private sector, is needed.

Professional transitioning will be among the new assignments for higher education.[14] Development of very special curricular material is needed. A remodeling of existing materials will not do. Required new knowledge and skills are defined by new job opportunities, but past personal experience and expertise must not be lost. One must also address the personal and emotional stress that accompanies forced career change. Rekindling the enthusiasm and motivation for learning can be a special and largely overlooked element, especially for the technically entrenched.

Professional transitioning is an educational process that enables a seasoned professional to rapidly and efficiently accommodate the changes required of a new career or new job responsibilities. The process must optimize the utilization of the individual's existing skills and capabilities while expediting, both in time and cost, the emotional, intuitive, and cognitive development prescribed for their new professional assignment. Given the rapid change of our technology, this task represents the single most important element in our country's move towards continuing and stable economic growth. We simply cannot lose or waste an already developed national intellectual capital.

Professional transitioning is addressed because it is one of the least discussed and least-developed elements in the economic development picture. Also, professional transitioning stands out alone as the enabling element for managing the creation of new ideas, innovation, and concepts that ultimately create both meaningful jobs and longer term job security.

While the concept of professional transitioning focuses on curriculum for the adult student, it is important to keep in mind that any career direction must be focused on areas where job opportunities are in high demand. The Cannon Group, led by Edward J. Cannon, researched trends in professional employment.[15] They analyzed existing literature and conducted a number of interviews with corporate higher management in order to determine the nature of new and emerging professional careers. The product of this study is a very revealing and surprising profile of comments from CEOs and upper management. A common theme seems to be that, as corporations release or "outplace" professionals, they do so without a very clear picture of how these individuals will adjust to their abrupt career shift. At the same time, in organizations that massively reassign their current personnel, there seems to be considerable uncertainty about the level and appropriateness of internal retraining that is required to maintain the technical and managerial long-term strength of the organization.

One of the primary characteristics of professional transitioning education is the ability to provide a "fast track" for well-educated professionals in an individualized learning format delivered at a convenient location. The advent of massive information networks and the birth of new techniques in knowledge-based systems will play a major role in the enabling of these new educational processes. Presbury and Benson[16] look carefully at some of the computer-based tools that have been used to create various levels of interactive instructional modules. Several of the most popular are discussed in detail. Several instructional development tools currently under beta test are also reviewed.

Drs. Presbury and Benson provide a remarkable insight into the variety of elements that must play a major role in the complex process of learning-teaching in professional transitioning. One sees the definition of the complexity of professional transitioning from the viewpoint of experts in psychology and education. The adult undergoing a major disruption of personal and professional stability requires motivational considerations of specialized character. In addition, the cognitive process itself needs to be refocused when dealing with the experience and paradigms of the professional. The adult student is likely to be directed towards greater breadths in a technical sense combined with new perspectives on problem definition and solution. All of this points to the conclusion that, quite likely, a new educational process will emerge for the professional. Complicating this problem is the fact that the nature of emerging job opportunities is extremely broad and, therefore, will require instructional curriculum of great variety. The consolidation and convergence of these ideas leads one to conclude that while the world is moving toward virtual companies and organizations, so too will many educational systems move toward greater adaptability and the virtual structure. Presbury and Benson present a remarkably lucid explanation of this complex educational requirement, drawing the conclusion that computerized systems and broad-based use of large knowledge-based networks provide the optimum tools to meet the requirements for the transitioning professional.

Economic development cannot be sustained without major attention to the subject of our country's professional strengths. The future will require new and innovative mixtures of many basic disciplines of the past. Our transition from the industrial to the knowledge era requires extraordinary readjustments in economic and social structures. However, knowledge-based systems can show us how to adjust the educational processes to support the rapid and comprehensive response of our professionals to the problems of the day.

Work needed to develop professional transitioning programs can draw heavily from ongoing implementation experience in integrated studies curriculum development, such as that being done at Marshall University in West Virginia and James Madison University in Virginia.

Educational change is a micro-movement process. New degree options and even new course sequences come after painstaking, time-consuming reviews. When society, economically and socially, could accommodate this schedule, the process was acceptable. However, presently, education must not be seen as the least responsive and slowest element in meeting the needs of our citizens.

It seems clear that we are moving towards the development of systems that must accommodate the time schedules, content, interests, and career directions of the learners. The need for professionals to cross the traditional disciplinary boundaries is now commonplace in every career track. Bankers must understand the world of electronic funds transfer, encryption of data, and smart loan analyzers. Corporations are crossing geographical and cultural boundaries in each day's work in order to assemble services or products for customers that may range from manufacturers of brackets to assemblers of packaging for pharmaceuticals. This kind of interdisciplinary traffic is commonplace in the job market but is not the traditional way of working on many university campuses.

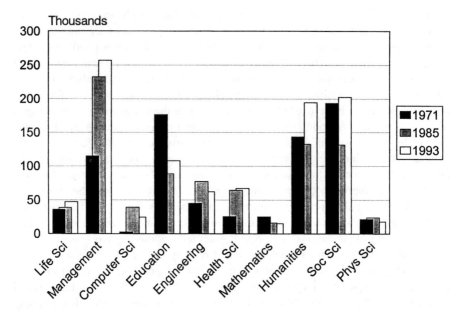

FIGURE 1.4. This data illustrates a two-decade shift in bachelor's degrees by majors. Note the rebounds from 1985 to 1993 in the humanities and social sciences against lack of activities in mathematics, physical sciences, and computer sciences. Integrated science and knowledge-based curricular redesign will likely produce another significant redistribution of these degrees in the near future. (From U.S. Department of Education, National Center for Educational Statistics, *Digest of Education Statistics,* 1995.)

A LOOK TO THE FUTURE

The changes in the job market, along with new career tracks, have led to a redistribution among the number of students graduating within the different majors offered in higher education (Figure 1.4). Startled by some of these unforeseen and significant shifts, universities are looking to their curricular offerings to accommodate the change taking place. In partnership with advances in computation, developments in knowledge-based systems appear as vital tools for implementing new options in the teaching-learning arena. Multimedia communication across the disciplines, with computers and networks playing a central important role, is encouraging student-to-student learning and faculty development.

The engineering profession is at the very center of the career and education revolution as representative of past and likely future changes in teaching-learning. Engineers live in the future. By experience and intuition, engineers envision today's design as an incremental move toward tomorrow's innovation. This vision ought to give an edge on predicting the future of the profession into the 21st century. Start by recognizing some of the fundamentals of engineering that have endured over the past 30 years or so. Most of these will persist over the next three decades. Significant fundamentals include the following.

- An adaptability for change in response to extraordinary technical and societal priorities
- Strong roots in modeling and simulation for both the analysis and synthesis of complex problems
- An understanding of and cooperation between the private sector, practicing engineers, and education
- A persistent desire to create effective verbal and visual communication

Components of these fundamentals that have been taught, learned, and practiced will trace a dynamic path into the 21st century. Herbert Hoover saw it this way:

> The great liability of the engineer compared to men of other professions is that his works are out in the open where all can see them. His acts, step by step, are in hard substance. He cannot argue them into thin air or blame the judges or lawyers. He cannot, like the politicians, screen his shortcomings by blaming his opponents and hope the people will forget. The engineer simply cannot deny he did it. If his works do not work, he is damned.

A look at how technology has evolved during the past 30 years may help identify some of the critical factors that could advance or severely limit the continuing development of technology, and society with it.

In 1965, a state-of-the-art multimedia instructional laboratory consisted of teletype/black and white monitor output devices for students interfaced to a host computer with 256K of memory. The instructor's station housed a black and white video camera, a 16-track analog recorder, a reel-to-reel black and white video recorder, and a half dozen carousel projectors for image memory.

Today's laboratory has docking stations for student laptop computers, high resolution color graphics, and virtually instant access to text, images, sound, and full motion video. Some students are already working at virtual reality terminals. Project this change 30 years into the future and invoke the imagination to conceive of a concurrent sensory environment delivering walk-through knowledge modules.

In 1965, a small host computer (DEC PDP-8) in the Clemson University Engineering Laboratories interconnected 17 high schools in South Carolina. The Internet, 30 years later, makes sophisticated global communications appear mundane. Will technology in 2025 engender a teacher-learner environment where people, place, and time rarely coincide?

In 1965, engineering students studied servomechanisms, steam generator design, and vacuum tube electronics. Now, we are designing orbiting nuclear power reactors, expert system technologies, global manufacturing, CAT scanning, and re-engineering the organization.

These reflections on the past suggest extraordinary change lies ahead. The best survival mechanism is to make a lifelong commitment to learning, including the recognition that learning will assume diverse constructs as a career advances. These constructs will be very different from routine classroom lectures offered on the university campus. Educators will find it necessary to continually adjust their teaching methods in order to prepare students for self-initiated instruction that will likely cross many currently "compartmentalized" disciplines. The present lecture, class-room, and lock-step instruction cannot survive the next 30 years.

The industrial era has yielded to the knowledge-based era, where advances in knowledge systems will expand at a stunning rate during the next 30 years. During the period 1965 to 1995, the necessary platforms for knowledge collection, storage, repackaging, and transmission were created. These are ready for exploitation at an unprecedented level, even in the near term.

Complementing the growth and development of knowledge-based systems is the communications platform that will be used to construct and deliver knowledge packages globally. For example, there are trial programs of technology in which subscribers can connect to TV programs over phone lines. Others are talking over coaxial TV cables and, of course, conversation over the Internet is here. Cryptog-raphy will play an important role in the future, but thanks to our past research and development from NSA, CIA, and other agencies, we still remain global leaders in this technology. Wireless, local area networks are being discussed, and of course World Wide Web pages may be pulled through twisted copper pairs using compres-sion and modulation techniques.

With the power of multimedia as an educational and information delivery device, the evolution of advanced electronic communications will receive continual pressure for upgrading. The work of Bergando and Davidson at AT&T Bell Laboratories have demonstrated experimental systems with extraordinary bandwidth.[17]

ISDN has received new emphasis. ATM (asynchronous transfer mode) networks are receiving heavy investments. In addition to all this, thoughts are given to the next generation Internet protocols. Early in 1997, Internet protocol IPv6 is expected to accommodate growth and capability in the multimedia requirements. Digital wireless has just begun deployment in the U.S. Regardless of who emerges with the best technology in service, it is bound to dramatically affect the way users access knowledge bases. The long sought for dream of educators is to have each of their students with a portable computer able to access the resources of the university from anywhere at any time. Courseware, laboratory experiments, student performance, linkages to benefactors and parents, are all in the near future for everyone.

The search continues for how to optimize and assemble knowledge units for the user while at the same time providing the network to deliver it, including innovations that make, through compression technology, old phone line twisted pairs look like fiber. All this work paints a bright picture for the future providing that we understand the basic and fundamental difference between the transmission of data and the packaging, structuring, and delivering of knowledge units.

REFERENCES

1. Bugliarello, G., The global generation, transmission and diffusion of knowledge and the needs of the less developed countries (LDCs), National Research Council World Bank Symposium, Nov. 1994.
2. Boulding, K., 1968.

3. Stewart, T., Brainpower, *Fortune*, June 3, 1991.
4. Scardamalia, M. and Bereiter, C., An architecture for collaborative knowledge building, in *Computer-Based Learning Environments and Problem Solving*, DeCort, E., Ed., Springer-Verlag, Berlin/Heidelberg, 1992.
5. Massy, W.F., New thinking on academic restructuring, *Priorities*, Winter 1996.
6. Bell, T.E., Adam, J.A., and Lowe, S.J., Communications, *IEEE Spectrum*, 33, 30, 1996.
7. Kennedy, R.C., Intranet explosion, *PCComputing*, 145, June 1996.
8. Report of the Blue-Ribbon Panel for the Proposed College of Applied Science and Technology, James Madison University, Harrisonburg, VA, 1990.
9. Spar, D. and Bussgang, J.J., The net, *Harvard Business Review*, 74, 125, 1996.
10. Armstrong, A. and Hagel, J., III, The real value of on-line communities, *Harvard Business Review*, 74, 134, 1996.
11. Wheatley, Margaret J., *Leadership and the New Science*, Berrett-Kochler Publishers, Inc., San Francisco, CA, 1992.
12. Smith, Hedrick, *Rethinking America*, Random House, New York, 1995.
13. White, R.M., A strategy for the national labs, *Technology Review*, 97, 69, 1994.
14. Wilcox, L.C., Engineering for the 21st century, *ASME Journal*, 1994.
15. Cannon, E.J., *Professional Transitioning with Expert Systems*, The Cannon Group, Tulsa, OK, 1993.
16. Presbury, J.A., Benson, J.A., and McKee, J.E., The mock turtle's lament: The cost of critical thinking, *Virginia Counselor's Journal*, 20, 12, 1992.
17. Bell, T.E., Adam, J.A., and Lowe, S.J., Communications, *IEEE Spectrum*, 33, 30, 1996.

2 Knowledge Management: Dealing Intelligently with Knowledge

Rob van der Spek and André Spijkervet

CONTENTS

0-8493-3116-1/97/$0.00+$.50

INTRODUCTION

As a member of a management team and/or as an employee of a professional organization, you are constantly confronted with questions about the role of knowledge in the company. These questions can vary from strategic questions to operational matters. The following are examples of such questions.

- What kind of knowledge do we actually have within the organization? Who else in the market has this knowledge? Which knowledge provides opportunities for developing new products in the short term? Do we have in-house knowledge with which the market can be changed in the long term?
- Which knowledge areas must we develop in the near future? Which must we develop ourselves and which will other organizations develop? Which knowledge will dominate the market in the next few years in the form of products and services?
- How are we going to develop this new knowledge? Must we work together with other organizations? Can we follow training courses? Who can develop this knowledge within the company?
- How can we transfer existing knowledge better and faster to colleagues and new employees? How can we make knowledge more accessible to others in the company? How can I document my own knowledge so that I can reuse it again?
- How can we ensure that we apply all available knowledge for producing a product? How can I ensure that I get a quick reply to a question with which I am struggling?

All of these questions have to do with the way in which we organize and guide the development and application of knowledge in a company and organization. At all levels of an organization, management has always been responsible for the effective allocation of people, resources, and tools, the planning of processes, and the evaluation of the results. In many organizations, however, answering the above questions has gradually assumed a central position in the daily tasks of managers. This is a role which demands a lot of the strategic insight, the problem solving capacity, and the tact of a manager.

WHAT IS MORE DISTINCTIVE THAN THE CAPACITY TO BUILD UP AND APPLY KNOWLEDGE OPTIMALLY?

In the past few years, a good deal has been written about the fact that companies are being forced to reconsider their business practices. Terms such as Business Process Re-engineering (BPR), process rationalization, Total Quality Management (TQM), and Learning Organizations were introduced. The various authors agree on the causes, namely:

- Increasing competition, in particular from the Asian continent, influenced by deregulation and new legislation
- Changing customer demands regarding time, cost, flexibility, and quality
- Fast rate of changes under the influence of technology, science, politics, and socioeconomic factors

Under various labels, many organizations and companies have started programs in which they attempt to face the above influences. It is not so surprising that more and more people are concluding that improving the generation and use of knowledge in the Western production processes is the only way to react to the above-mentioned developments. Best-sellers from the management literature report on developments such as: "The rise of the expert company" (Feigenbaum), "The knowledge society" (Drücker), "The intelligent organisation" (Quinn), and 'The knowledge-creating company" (Nonaka). Various developments are relevant in this context:

- The *knowledge intensity* of products and services is increasing rapidly which is mainly reflected in the cost structure
- The knowledge which is required for implementing business processes changes *quicker* as a result of technological and scientific developments and changing economic relations
- The pressure of *time under which* decisions must be taken is increasing
- The *mobility of professionals* is increasing through changing labor relations (e.g., increasing number of freelancers) and technological possibilities (e.g., tele-working); strategic knowledge can thereby "seep away" to competitors. Moreover, knowledge is increasingly being hired on a worldwide market.

Not only are authors noting that the effectiveness of the knowledge household is crucial to achieve a permanent competitive advantage, but that the efficiency of knowledge-intensive core processes must be increased to meet the demands of cost reduction. It is therefore not only a matter of applying the right knowledge at the right place and at the right time, but also at a minimal cost.

In the past few years, organizations have devoted a lot of attention and money on improving the supply of information. In many cases, it was forgotten that making qualitatively high-level decisions requires both knowledge and data. These must be well-adjusted to each other.

These developments force organizations to make a greater effort to enhance the application and development of knowledge, thereby improving their competitive position. Organizations that are not capable of doing this develop all kinds of bottlenecks/pathologies, often with far-reaching consequences.

BOTTLENECKS ENCOUNTERED

Some of the bottlenecks mentioned by companies which are caused by an inadequate knowledge household include the following.

- It takes too long before new knowledge is optimally applied by the whole company.
- Learning does take place within the company but at a slow rate while the environment is learning at a faster rate.
- In order to generate an end-product, a large (often informal) network of people is needed, consulting together.
- Strategic knowledge seeps away through retirement, reorganization, project-based work, job rotation, and shift work.
- The same knowledge is developed anew because knowledge is not recorded, because it is not known which knowledge is present, or because it is not known who within the company has the required knowledge.
- Unnecessary loss of time occurs during handling because knowledge is not effectively organized; often knowledge is not present at the interface between the customer and the company (making quotes, advising, customer service).
- Mistakes are made due to inadequate knowledge which increases cost of maintenance, after-sales, or replacement which directly affects the competitive position.
- Where knowledge is recorded, it is often only "know how" and not "know where" or "know why."
- Employees become frustrated because knowledge is not accessible within the company.
- There is insufficient investment in knowledge areas which will ultimately develop new markets.

Companies are, however, often not aware of the fact that the standard symptoms derive from a poor knowledge household. These symptoms include:

- Complaints of external customers about quality
- Complaints of internal customers about quality

- High costs of a process
- Long production periods

KNOWLEDGE MANAGEMENT

The importance of knowledge to the continuity of companies and organizations is self-evident. One may therefore expect that the control and management of knowledge and processes relating to knowledge is always well organized. In reality, this is often not the case. This is indicated by, among other things, the number of bottlenecks which are described during interviews and surveys. The introduction of the concept of knowledge management is therefore no luxury, but a pure necessity.

Knowledge management aims to provide instruments to employees of professional organizations who are confronted with the need to optimize the control and management of their most crucial production factor. Knowledge management aims at *preventing* bottlenecks and cashing in on opportunities by determining from a *strategic perspective* which requirements will be set for the knowledge household in the future. Knowledge management will also focus on solving existing bottlenecks and therefore is strongly *problem-oriented.*

The core of knowledge management is the *organization of processes* in which

- new knowledge is developed,
- knowledge is distributed to those who need it,
- knowledge is made accessible for both future use and use by the whole organization, and
- knowledge areas are combined.

Research has shown that in many organizations hardly any structural attention is paid to knowledge management. Of course there are all kinds of activities in subfields, and some already have a long history, such as, the activities of training officers and Human Resource Managers. What is often lacking, however, is coordination between various activities and departments, which means that synergy rarely occurs between various initiatives.

Improvements will ultimately be about adapting the behavior, ways of work, and, in particular, habits of people. The culture of an organization, behavioral characteristics of people, and existing coalitions have, in reality, consequences for the feasibility of improvement actions. These factors will therefore also play a prominent role in the discussion on instruments for the improvement of knowledge intensive organizations.

Survey at 60 Dutch organizations and companies

In 1994, the Knowledge Management Network, a joint initiative by companies and nonprofit organizations, held a survey among companies and knowledge institutions with the aim of examining how managers think about the role of knowledge as production factor, the effectiveness of the current knowledge policy, the bottlenecks regarding the application of knowledge, and the role of management in establishing objectives and conditions for knowledge management.

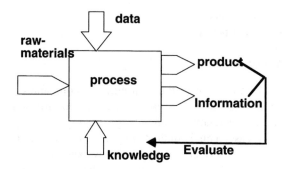

FIGURE 2.1   Processes and knowledge.

49 respondents from 42 selected profit-organizations and 31 respondents from 18 nonprofit organizations participated in the survey.

The survey showed that the majority of the organizations has had problems with the availability of vital knowledge and expects similar situations in the near future. A number of surprising results include:

52% of the companies encounter problems in transferring knowledge when restructuring processes and transferring personnel.

57% of the respondents report that costly mistakes were made through not having the right knowledge at the right place and at the right time.

80% of all respondents report situations in which only 1 or 2 people have certain crucial expertise.

KNOWLEDGE IN ORGANIZATIONS

Business processes form the core of a company or organization. Based on a projected or real customer demand, a product or service is delivered through a chain of process stages. Within each business process, knowledge is used which is present in people and in the form of other knowledge bearers such as electronic media, computer systems, paper media, and machines (Figure 2.1).

The added value of the final product is determined to a large extent by the quality of the knowledge applied. This quality is in turn determined by the quality of the knowledge bearers involved and in particular their interrelations.

Knowledge is the whole set of insights, experiences, and procedures which are considered correct and true and which therefore guide the thoughts, behavior, and communication of people. Knowledge is always applicable in several situations and over a relatively long period of time.

A complementary definition of knowledge is (Gardner 1995)

- knowing which information is needed ("know what"),
- knowing how information must be processed ("know how"),
- knowing why which information is needed ("know why"),
- knowing where information can be found to achieve a specific result ("know where"), and
- knowing when which information is needed ("know when").

Data are symbols which have not yet been interpreted. We are confronted daily with data in various forms. Examples are a red light on a dashboard or a set of process data in a factory.

Information is data which has been assigned a meaning. A chauffeur assigns meaning to the red light and will stop because, according to his interpretation, over-heating has occurred. A graph provides information on the relation between aspects on the horizontal and vertical axis of the graph and shows, e.g., that there is a certain trend. Information is always linked to a specific situation and has only a limited validity.

Knowledge is that which enables people to assign a meaning to data and thereby generate information.

Knowledge, therefore, enables people to act and to intelligently deal with all the information sources available. A red light on a dashboard can mean a low oil level, a low petrol level, or a warning that a brake is not functioning. Knowledge about the car is therefore required to choose the right action or ask for the right information! This action component is an essential aspect of knowledge.

Knowledge is company specific. A transport company, for example, that wants to serve its clients as cheaply as possible will develop different knowledge to a company that wants to serve its clients within a certain time frame. The generation and appli-cation of knowledge is determined by the mission and objectives of the company.

An important feature of knowledge is that by using it in processes and by acting in the environment, people are able to adapt this knowledge. This learning process sometimes occurs consciously, but it is usually unconscious.

STRATEGIC IMPORTANCE OF KNOWLEDGE

Within organizations, not all knowledge plays an equal role. A distinction can be made between different knowledge areas depending on the strategic importance to the organization, the growth potential, and the stage of development of knowledge areas. This approach is analogous to the way in which, for example, strategies are formulated in a portfolio-analysis (e.g., Boston Consulting Group). The life cycle of knowledge areas describes the rise, maturity, and decline of knowledge areas (Figure 2.2).

The classes of knowledge areas which are distinguished are

- *Promising knowledge areas*
  These are still in their infancy but have demonstrated to have the potential to radically change the execution of one or more tasks of an organization.
- *Key ("core") knowledge areas*
  These distinguish the organization from other companies. They have the greatest influence on the unique position of the organization. Hamel and Prahalad call these the *core competences*.
- *Basic knowledge areas*
  These are essential/necessary for carrying out the activities of an organi-zation. This knowledge is widely available in all similar organizations.
- *Outdated knowledge areas*
  These are hardly applied any more in business processes.

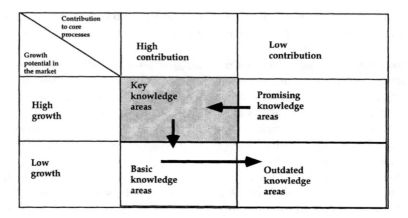

FIGURE 2.2   Knowledge life-cycle.

People develop ideas in all kinds of situations, and some ideas develop into promising knowledge areas. Some of these develop, under suitable conditions, into the key knowledge areas of a company. Through diffusion, these are gradually applied more widely in a certain sector, whereby the distinctive capacity which can be achieved on the basis of this knowledge decreases. Finally, knowledge will become outdated and it is no longer desirable that this knowledge is applied in the business processes.

Critical knowledge areas are those knowledge areas which are of vital importance for the prosperity of the company now and in the future. Examples of critical knowledge areas are knowledge areas where in the next few years (using the explicit control and management of knowledge):

- there are opportunities to achieve a significant improvement in efficiency and/or effectiveness,
- there are opportunities to enter new markets, or
- it is possible to anticipate (possible) events with major negative effects (e.g., the departure of the only expert within an important knowledge area)

Both basic-, core-, as well as promising knowledge areas can of course be critical to the competitive capacity of an organization over a longer period.

Not only is expert knowledge which is applied in the production process critical for the company, but also knowledge *about the company itself* and knowledge about the *market and relevant external developments* is also critical.

KNOWLEDGE ABOUT THE COMPANY

First, knowledge about the company is critical for a proper coordination between the primary activities. A salesman, for example, must have knowledge about the production process in order to plan realistic delivery dates.

Second, knowledge about the company is critical for carrying out the supporting processes and management activities.

Ask yourself, as an executive, the following questions:

- What is the mission of the company?
- Are the objectives known and operationalized down to the lowest level?
- What is the structure of the company?
- Why do we produce in this way?
- Who is who in the company?
- What are the informal relations between the employees?
- How do we actually learn in this company?
- How is our knowledge household built up?

We are dealing here with knowledge which can only be built up while working in the company. You will probably realize that this is crucial knowledge for making both policy decisions as well as solving daily problems. The knowledge is incidentally not only important to the executives but also to the employees on the shop floor! It is important that when production problems occur, the people involved know who must be alerted to adequately resolve the problems. New employees often do have enough formal knowledge, but still lack this company-specific knowledge.

KNOWLEDGE ABOUT THE MARKET

Besides knowledge which is applied in the primary processes and knowledge about the company itself, a company must, of course, also have knowledge about the markets in which the products will be sold. This knowledge is of particular importance for marketing activities, but also for determining strategic policy. This knowledge covers current and potential markets, customer profiles, characteristics of customers, competitors, external developments which may have influence (e.g., changes in legislation and regulation, demographic developments, political and economic factors). Recently, a lot of attention has been paid to the phenomenon "Business Intelligence." This refers in particular to the competence of a company to analyze the market and transform soft and hard information into knowledge about the position of the company and the competition.

DIMENSIONS OF THE KNOWLEDGE HOUSEHOLD

Two different dimensions can be distinguished in the design of the knowledge household. In the first place, the *processes* in which the management activities regarding knowledge are carried out. In the second place, the *structure* of the knowledge household in which the bearers of knowledge, their specific characteristics and their mutual relations are contained. Both dimensions will be discussed.

BASIC OPERATIONS ON KNOWLEDGE

In the literature, four processes are distinguished in which basic operations for knowledge management are implemented (Wiig, 1991; Nonaka, 1992; van den Broeck, 1994). These processes are represented in Figure 2.3.

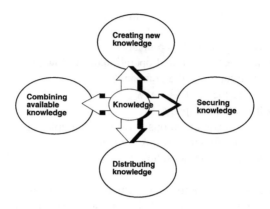

FIGURE 2.3   Four basic operations on knowledge.

These basic processes can be described as follows:

- *Developing new knowledge*
  Companies survive by developing new knowledge based on creative ideas, the analysis of mistakes, daily experiences, and hard work in research and development departments.
- *Securing new and existing knowledge*
  Knowledge which has an individual basis must be made as accessible as possible for the collective, and available in the right place, at the right time for the company.
- *Distributing knowledge*
  Knowledge must be actively distributed to those who can make use of it. The turn-around speed of knowledge in a company is increasingly crucial to the business processes. Distribution of knowledge can, of course, be achieved by several methods — by transferring people, by organizing courses, by giving presentations or through an internal video-news show.
- *Combining available knowledge*
  A company can only perform at its best if all available knowledge areas are combined. Products and services are increasingly being developed by multidisciplinary teams.

Bottlenecks and Basic Operations on Knowledge

The following bottlenecks have been formulated in terms of the processes linked to the management of knowledge.

- **Developing new knowledge**
- Not enough is learned from the developments in the market. Knowledge about potential markets, current markets, and existing or new competitors is not structurally developed.

- The structural development of new ideas into market-ready products does not take place. Often because there is not enough patience and commitment to give new ideas a chance. In many cases, there is also no clear vision of the future, so that it is actually impossible to determine which ideas must be worked out.
- **Securing new and existing knowledge**
- Implicit knowledge is not recorded and individual learning processes are not transferred to a collective learning process.
- Knowledge which is recorded is often not traceable and is inaccessible.
- Individual interpretations lead to a drop in quality of the product.
- **Distributing knowledge**
- It takes too long before new knowledge is actually applied in all the places where it is required.
- It takes too long before new employees have built up enough knowledge.
- Knowledge which was developed on the work floor is often not passed on to colleagues (e.g., in shift work, project employees).
- **Combining available knowledge**
- Knowledge is not combined because people often do not know who has which knowledge.
- People often do not know which knowledge is needed for producing an optimal product or service.
- People from different knowledge areas often do not communicate well with each other due to a lack of a common vocabulary.

STRUCTURE OF THE KNOWLEDGE HOUSEHOLD

A number of structural features of knowledge can be distinguished (see Figure 2.4):

The *form* of knowledge denotes the medium or bearer in which knowledge is stored, in the first place, of course, people. People are active bearers of knowledge, which means that they are capable of applying, developing, and increasing knowledge in their daily experiences. A characteristic feature of knowledge as a raw material is that it does not decrease but rather increases in size! People are often not aware of the knowledge which they apply. In this context, a distinction is sometimes made between "tacit" and "explicit" knowledge.

Knowledge is also present in books, memos, manuals, and other documentation. A characteristic feature of these knowledge bearers is that they are passive. This means that they are not capable of either independently applying or developing knowledge. Knowledge in a written form must always be applied and updated by people. This encourages the description of information bearers, instead of knowledge bearers, for these mediums.

Technology and in particular computer technology assumes an intermediate position. Devices and automated systems are capable of independently carrying out tasks using the knowledge and information that is stored internally. Software makes it possible to program behavior of systems based on the knowledge which designers consider necessary. The rise of "intelligent software" plays an important role here.

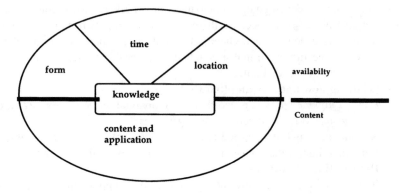

FIGURE 2.4   Four features of knowledge.

Certain computer systems, also called knowledge-based systems, are increasingly becoming active bearers of knowledge that can independently apply this knowledge. Examples are making a diagnosis, compiling advice, etc. Computer systems are increasingly able to generate knowledge based on input data.

The *location* of knowledge describes the position of the knowledge bearers within the company or organization. Knowledge can be localized in the front-office, or in the back-office, or on the other side of the world; paper bearers can be made centrally available in a library or to employees in their own rooms.

The *dimension of time* describes the availability, over time, of knowledge and the temporal aspects which are related to the use of the knowledge. People can be on call 24 hours a day or only a certain part of the day. In some fields of work, humans cannot react quickly enough to the amount of data and the speed at which situations can change, which makes it necessary to use computers. This applies, for example, to warfare, but also to detecting fraud in credit card transactions.

The dimension of *content* describes both the procedures and experiences in a work domain as well as the way they can be applied.

Bottlenecks and Features of Knowledge

Considering knowledge to have four structural features is a means of clustering causes for problems occuring in these core areas. In reality, bottlenecks will often have combined causes. Causes from the perspective of the four dimensions of knowledge include the following.

Form

Knowledge is not present in the most optimal form, e.g., knowledge is only present in one person or is hidden in thick manuals.

The form in which knowledge is stored is not suitable for maintenance, which is why knowledge is not maintained or only at a very high cost.

Content

Knowledge is not complete, e.g., because knowledge is only applied from one discipline.

Knowledge is not current, e.g., knowledge is not adapted to changing circumstances ("no learning occurs").

Knowledge is not uniform, e.g., decision-makers use personal interpretations which leads to different results.

Time

Knowledge is not available at the time it is needed, e.g., because the expert is only present between 10:00 and 11:00 at the office.

Location

Knowledge is not present at the place where the business process is carried out, e.g., knowledge is only present at the head office but is particularly needed at branch offices. Knowledge can also be spread across an organization so that synergy does not occur.

Availability of Knowledge

One rule of thumb from the guru of the Business Process Re-engineering movement, Michael Hammer, says "Make complete handling of the business process possible where the customer comes into contact with the company." Michael Hammer, in: Restructuring Business Processes: In *Harvard Holland Review*, 27, 7 t/m 15 (1991). Actual implementation of this rule of thumb can only occur if all the knowledge is also available where the business process must be dealt with. If this is not the case, it inevitably leads to delay.

KNOWLEDGE MANAGEMENT

Knowledge management is therefore the explicit control and management of knowledge within an organization aimed at achieving the company's objectives.

Knowledge management entails:

- formulating a strategic policy for the development and application of knowledge,
- executing the knowledge policy with the support of all parties within the organization, and
- improving the organization where knowledge is not optimally used or is not adapted to changing circumstances.

The added-value of the concept of knowledge management is mainly that it is a perspective in which improvements are sought across the borders of the "traditional" management functions of knowledge bearers such as personnel management, training, and documentation management. KM aims at improving the performance of processes, organizations, and systems in general from the perspective that knowledge is the crucial production factor.

Moreover, KM aims at an integration of strategy formation and executive tasks where learning about the application and development of knowledge assumes a central role. The control and management of the life cycles of knowledge areas in organizations forms a crucial aspect of strategic policy in organizations. In this context, it also applies that: "It is better to prevent than to cure." Only if a company is able to adjust various life cycles of knowledge areas and to allocate materials, people, and tools so that life cycles can enhance each other, can we speak of structural innovation.

KM must be considered as an integral part of management tasks. It is not expected that there will be "knowledge managers." In principle, every manager must control and manage the aspect of knowledge as part of his/her daily work.

There is no cut-and-dried approach for KM. For each organization, the approach, the methods and techniques, and the instruments used will differ. Ultimately, it is not important how KM is modeled and with which tools, just as long as it happens.

KNOWLEDGE MANAGEMENT OBJECTIVES

The objectives of knowledge management can be formulated in terms of both the processes which are a part of knowledge management as well as in terms of the structure of knowledge.

The following objectives can be set in terms of the processes within the context of knowledge management:

- Ensure an effective and efficient *development of new knowledge* and improvement of existing knowledge with a view to the strategy of the organization and the individual objectives of the employees.
- Ensure a specific *distribution of new knowledge* to other departments and *transfer of knowledge* to new employees through knowledge transfer or relocation of knowledge bearers.
- Ensure an *effective securing of knowledge* which is also easily accessible to the whole organization.
- Ensure the *effective and efficient combination* of the best knowledge available within a company or network of companies.

In terms of the dimensions of knowledge, the following objectives can be set using knowledge management:

- Keep the *content* of knowledge bearers up to date and correct under changing circumstances; apply the best knowledge.
- Make the *location* of knowledge bearers optimal in the context of business processes; apply knowledge at the best location.
- Improve the *form* of knowledge bearers in relation to the users and the expected use of it; apply knowledge in the best form.
- Adapt the *availability* of knowledge to the time that the knowledge is needed. For example, consider the availability of knowledge when this knowledge is needed within the context of a business process; apply knowledge when required.

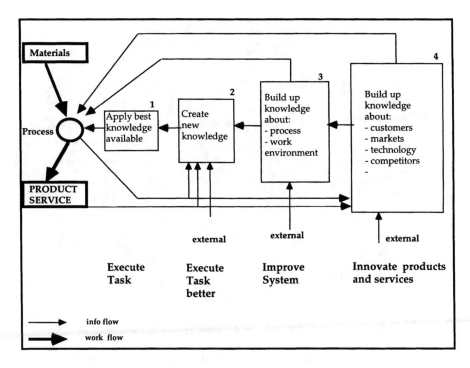

FIGURE 2.5   Levels of ambition of knowledge management.

LEVELS OF AMBITION OF KNOWLEDGE MANAGEMENT

Based on Deming's work in the field of quality management, Wiig distinguishes four levels of ambition for improving knowledge (see Figure 2.5).

At level 1, the focus is on applying the best available knowledge for executing a task. At level 2, new knowledge is built up with the aim of improving the execution of a task. Level 1 and 2 are very similar to the "single loop learning" as described by Argyris. At level 3, knowledge *about* the process and its properties is built up in order to improve the system. This level closely resembles what Argyris calls "double loop learning." Finally, at level 4 knowledge is built up about the market, the competitors, and the position of the company in its environment. The aim is then the innovation of the products and, perhaps, even the market.

Building Up Knowledge About Knowledge

In many publications, the ideal-type organization is discussed: the learning organization. In fact, a learning organization is an organization which is structurally capable of evaluating the results of knowledge-intensive work processes, adapting the knowledge, and applying the new knowledge quickly. Learning in an intelligent organization takes place consciously. It is very focused on collective learning and the deliberate improvement of the learning capacity. That means specifically that not only is new

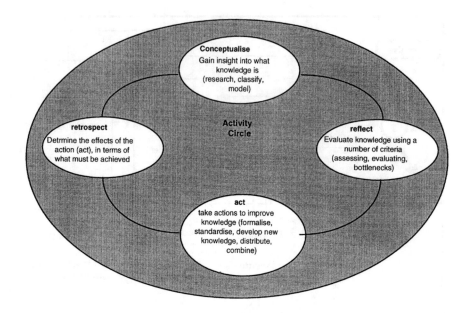

FIGURE 2.6   Conceptual model for knowledge management activities.

knowledge developed, but also that people think about the way in which this development takes place, how the new knowledge is distributed, which aspects play a role, etc. The organization of the learning process itself is therefore the subject of management activities! Knowledge management thereby supports the improvement of the learning capacity of organizations and is therefore closely linked to the idea of learning organizations. Organizing the learning process is an important part of KM.

PROBLEM APPROACH AS CONCEPTUAL MODEL FOR KNOWLEDGE MANAGEMENT

Although KM is different to other management activities in that it focuses on the aspect of knowledge, in reality it encompasses activities which fit within a more general problem-solving cycle (Argyris).

Specific interpretation for KM-activities generates the conceptual model as represented in Figure 2.6.

The KM activities ultimately control the basic operations on knowledge and the concrete knowledge bearers and deliver the following products:

- *Objectives* for knowledge development and application
- *Assessment of risks* regarding improvement actions
- *Conditions* for improvement actions
- *Instruments* for achieving the objectives set
- *Criteria* for measuring the performance of business processes and specific knowledge bearers

This conceptual model is very suitable for the KM-concept because:

- *KM is a learning process just like all management activities.*
  The learning process has been made explicit in this conceptual model. Through a repetition of conceptualizing, reflecting, acting, and assessment, it is possible to adapt the results of KM activities to changes in the environment, changing insights, and the changes which KM activities cause in the organization (intended but certainly also unintended!). The motto here is "Changes happen anyway; the trick is not to ignore them but, rather, to anticipate them."
- *It guides the structuring of activities and the required methods and techniques.*
  Modeling techniques, for instance, are specifically applicable in the conceptualization phase whereas concrete approaches for the realization of new knowledge bearers are specifically applicable in the action phase.
- *It is applicable at all levels of the organization.*
  KM is a management activity which will be specifically interpreted at all levels of an organization.
  At a *strategic* level, global objectives for the knowledge policy are set, which must be attained in the long term, and general conditions for knowledge management are established.
  At a *tactical* level, a specific interpretation of concrete medium term improvement actions is given. Objectives will be further specified and the application of instruments will be detailed.
  Finally, at an *operational* level, specific improvement actions are effected. At all levels, however, *all* activities in the problem-solving cycle are present.
- *It is applicable to various situations.*
  KM can, for example, focus on the performance of a single work process, the functioning of a department or group. In addition to focused application, knowledge management is applicable in the context of the whole organization and also across several organizations.

TECHNIQUES AND INSTRUMENTS

The conceptual model described for knowledge management has a number of phases in which specific techniques and instruments can be applied. This is not the place to give a comprehensive account of methods and techniques. In Appendix 2.1 a global overview is given of the possible activities and products, based on the various phases.

Knowledge Audits Provide a Picture of the "Fitness" of Knowledge-Intensive Work Processes

During a "knowledge audit," one examines how the various dimensions of knowledge influence the critical customer demands, time, cost, quality, and flexibility. This can be done, for example, by following a chain of business processes and analyzing how the application and quality of knowledge influences the final result. At present, for example, a workshop is being organized by the "InnovatieCentrum Midden-Nederland" in which 12 medium-sized companies, supported by the Knowledge management Network, are independently carrying out an audit on the quality of the knowledge household. A number of checklists were developed for this purpose. Filling in the checklist is actually only an aid here. The main aim is to systematically examine a number of issues. The objective is the discussion which will hopefully develop within companies.

Interfaces Between Knowledge-Intensive Business Processes

A crucial point to consider when identifying causes of bottlenecks in the knowledge household are the *interfaces* between knowledge bearers. In business processes, different persons, departments, or even organizations work together to produce a product. Bad results are often due to an inadequate exchange of information so that the knowledge present cannot be optimally applied and new knowledge can not be built up. A better insight into the knowledge which is, or should be, present within cooperative processes, enables the improvement of the communication between actors. The central principle of this approach is therefore that the optimal application and development of knowledge is paramount and directs the way in which information is exchanged and is stored. *In this approach, information policy is therefore a derivative of knowledge policy.*

DIFFERENT APPROACHES

In the publications on the subject of knowledge management, a broad range of approaches is presented, each with their own emphasis on problems and solutions. The two most prominent approaches might seem at first sight to be "competitors."

First, there is a *system-oriented approach* which attempts to gain a better insight into supply and demand of knowledge and the quality of the organization as a "knowledge system." It does this by analyzing and documenting processes, actors, knowledge bearers, knowledge fields, and the dynamics in the work field. On the basis of such an analysis, bottlenecks and effects are identified. The characteristic feature of this approach is that knowledge is considered as a production factor which can be analyzed in isolation from the current bearers of the knowledge. Several options are then available for attaining improvements. This approach resembles theories like Business Process Re-engineering and quality management, as well as the "soft systems" approach which underlies theories on learning organizations.

Second, there is the approach which focuses on the improvement of professional organizations based on *people's behavioral criteria* and the cultural context. The independent professional is the central figure here. Improvement actions are not so much based on an analysis of the knowledge as on an abstract concept. The emphasis is much more on facilitating professionals so that they can apply their knowledge to the advantage of the organization and that their knowledge will be updated to ensure that this is also possible in the future. From this perspective, knowledge is

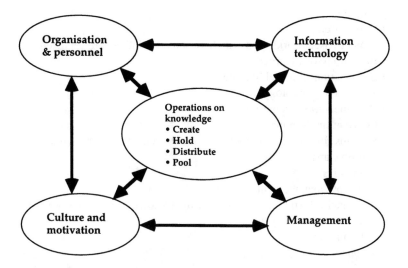

FIGURE 2.7 Different components of the structure of operations on knowledge.

considered as inseparable from humans. An extreme position within this approach states that making knowledge bearers and knowledge areas explicit in relation to business processes will lead to rigidity and is mainly a means for the management to get more grip on the employees. In this approach, several solution options are possible.

Within the Knowledge Management Network, there is also an active discussion on this topic, where the most plausible conclusion at present is that the truth lies somewhere in the middle.

BROAD RANGE OF INSTRUMENTS FOR IMPROVEMENT ACTIONS

KM has a broad range of instruments at its disposal which can be can be applied during improvement actions.

The organization of the management of knowledge is particularly determined by components such as culture, motivation of employees, organization, management, and information technology (Figure 2.7).

In general, three groups of instruments can be distinguished as follows.

1. Management, culture, and personnel
   • Strategy-development
   • Education and training
   • Recruitment and selection
   • Reward models
   • Adjusting management style
2. Organizational adjustments
   • Redesign of business processes
   • Adjusting control model

- Mergers
- "Outsourcing"
- Project-based working in which various disciplines are represented
- Bureau lessons learned
- Introduction of a buddy system
3. Information technology
   - Documentation technology
   - Information systems
   - Systems for supporting cooperation between persons/departments (Groupware)
   - Telematics
   - Workflow management systems
   - Personnel information system in which knowledge profiles are stored
   - Knowledge-based systems
   - Data mining
   - Intranets

Which instruments are most suitable depends on a complex set of factors, including the features of the knowledge bearers involved, the specific bottlenecks, and the environment in which the company operates. The major factor, however, is the corporate philosophy which sets objectives and conditions regarding the application of instruments. This covers aspects such as dealing with employees, the vision with regard to technology, the social responsibilities, etc.

Rewarding Knowledge Sharing

How many organizations are in fact there where the sharing of knowledge is rewarded? Employees are either rewarded without any evaluation of their functioning, or else on the basis of the individual results achieved. The emphasis on results forces employees and business units to shield off their knowledge in order to maintain their competitive position. Therefore, it is not a good basis for a better organization of the knowledge household.

KNOWLEDGE MANAGEMENT REQUIRES A MULTIDISCIPLINARY APPROACH

The nature of the intended activities makes it clear that KM is a strongly multidisciplinary approach. Different disciplines such as business economics, human resource management, organizational psychology, communication science, computer science, and operations research can make a contribution here. This does not only include instruments for achieving improvement actions, but also methods and techniques for understanding knowledge-intensive work processes and tracing causes for bottlenecks.

Several projects have also demonstrated that the added value of the KM approach for organizations lies in particular in the fact that the focus is on knowledge and not so much on specific methods and techniques from a single discipline.

BENCHMARKS ARE NECESSARY

In the context of KM activities, there is a direct need for benchmarks for measuring the performance of knowledge-intensive work processes, both for the analysis of the current situation ("How well are we doing?") as well as for the evaluation of improvement actions ("How well are we doing now compared to before?"). It is obvious that KM uses approaches here which have proved themselves in quality assurance.

A weak point in the KM approach is the analysis of financial aspects of knowledge. There are actually no methods available with which knowledge can be evaluated in financial-economic terms. However, if knowledge is considered as a production factor, it is plausible that knowledge will somehow figure on the accounts. A remarkable development in this context is the so-called "techno-lease" construction which Philips and Fokker have arranged with the RABO-bank.

In the magazine *Fortune* of October, 1994, a cover story is devoted to companies such as Canadian Imperial Bank of Commerce, DOW chemical, Hughes Space & Communications, and Skandia, which attempt to assign economic value to knowledge. The most important conclusion is actually that no one really knows how to do it, but that everyone is convinced that it is important. Companies are also not inclined to publish information about such a crucial business factor, which means that research in this field is having trouble taking off.

An honest conclusion must therefore be that a lot of work still has to be done here.

KNOWLEDGE MANAGEMENT REQUIRES AN IMAGE OF THE FUTURE

We have already indicated that KM is both reactive and proactive, and that it is not enough to know what we want to achieve *now*, but also what we want to achieve in the future. A proactive stance of KM requires that we develop a vision about which knowledge areas must be developed (Which promising knowledge areas will become core areas?), which knowledge areas must be examined (Which knowledge areas are promising?), but also which basic knowledge areas must be maintained or perhaps even rejected.

A crucial question which can be posed here is how organizations can predict the life cycle of knowledge areas. It is natural to look for solutions in the direction of scenario models. We must bear in mind here that the modern society is changing at such a pace that predictions are often only wild guesses. An interesting development in this context is that KM is also applied to knowledge which underlies predictions.

Knowledge about the Future

One of the most renowned scenario theorists, Peter Schwartz, has the following to say about this:

Articulating your mind-set—people often do not realise that their decision agendas are usually unconscious. Thus, the first step of the scenario process is making it conscious.

You begin by examining the mind sets which you personally use — consciously or unconsciously — to make judgements about the future. Think of this process as a form of research. Instead of gathering information out in the world, you gather information from within yourself. (Peter Schwartz: *The Long View,* page 53.)

KNOWLEDGE MANAGEMENT AS A CONTINUOUS LEARNING PROCESS IN A TURBULENT ENVIRONMENT

On the basis of the components discussed, the following framework emerges for knowledge management. This is depicted in Figure 2.8.

The core of KM is a learning process of conceptualizing, assessing, acting on, and evaluating alternates in cycles.

What is specific for KM is the focus on knowledge, knowledge bearers, and the organization of the processes which manage knowledge. It covers all activities that ensure an optimal control of factors which affect the functioning of the knowledge household.

Each phase of the learning process applies specific instruments to achieve the desired objectives. These objectives and the way in which they can be achieved can in fact be considered as the knowledge areas of knowledge management.

Office Lessons Learned of the Royal Army Corps (KL)
In June 1993, the KL set up an Office Lessons Learned. Major Rondel commented on the objective of this office: The Office Lessons Learned tries to prevent situations whereby peace-keeping missions have to keep reinventing the wheel. There is, for example, spread over the whole country, in different units, a good deal of experience. This experience was generated through evaluations of exercises which the units have done. Unfortunately, this information often remained at the battalion or brigade level, meaning that other units could not profit from it. Nowadays these experiences are analysed by our office and we then attempt to extract "Lessons Learned" from them: Practical facts that indicate how something can be done better or how it should not be done. (Source: LLC Courant, Butt 1994)

Giving recipes for carrying out the activities in each phase is practically impossible because this is determined by unique internal and external factors. For instance, in the context of the conceptualization phase, we could ask the question: is it always necessary to categorize which knowledge we already have before we can determine which knowledge is needed? Can we always make a systematic inventory considering the time and money at our disposal? And when we are ready with the inventory; has our competitor not already started developing new knowledge?

Example of a specific approach to knowledge development
Nedap attempts to promote the exchange of knowledge between the 430 employees as much as possible. CEO Westendorp says:

"It is important that people do not limit themselves to the boundaries of their function, but can see the whole picture. That promotes creativity. That is why people in the five product groups, the sales, and development and production groups cooperage intensively. Anyone who has an idea, or wants to develop more knowledge in a certain field,

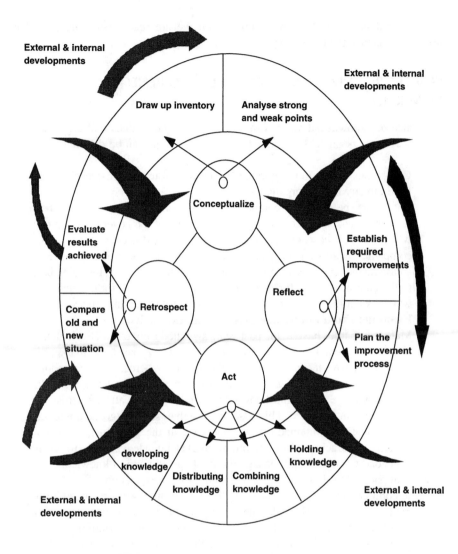

External & internal developments

External & internal developments

Draw up inventory

Analyse strong and weak points

Conceptualize

Evaluate results achieved

Establish required improvements

Reflect

Compare old and new situation

Retrospect

Plan the improvement process

Act

developing knowledge

Holding knowledge

Distributing knowledge

Combining knowledge

External & internal developments

External & internal developments

FIGURE 2.8   Framework for knowledge management.

is allowed all the space required. What does or does not belong to our core activities right now, is not relevant. The only guiding principle is that you have the feeling that the company may need it."

Westendorp describes the knowledge development process of NEDAP as controlled coincidence. "It is based on feelings, but is practically guided by your own background, the trade fairs you visit, the customers you talk to."

(CEO Westendorp of Nedap in Management team 4-4-94)

The learning process will ultimately be strongly affected by external developments, opportunities, and threats, which can occur at any moment in the learning process. The fact that we are enthusiastically taking stock does not mean that the outside world is sitting still and waiting until we are ready. But who said that knowledge management was easy?

WHICH ORGANIZATIONS CAN EXPECT TO BENEFIT FROM KM?

Experience has shown that various motives exist for companies and organizations to seriously embark on KM. A number of general groups can be distinguished:

1. Companies which must reorganize due to various circumstances or that have already been reorganized
   These may be industrial companies whose competitive position has been seriously weakened but also, for example, parts of the government system, which must cut costs or which have to privatize. KM is considered by such companies as a means of gaining more insight into the critical knowledge areas and the employees who have this knowledge. The need for going back to the core business is for many organizations the reason for examining to which extent knowledge management can contribute to this process.
   Example: From knowledge-driven to market-driven
   Many companies have in the past years switched from product-orientation to market orientation. Business units were set up based on markets in order to improve the connections to the market. Whereas these companies were previously organized around products and the related technologies/knowledge areas, the knowledge is now often scattered across the various business units. This also means that knowledge is no longer automatically maintained, as was the case in the past. In areas where benefits were achieved in combining market knowledge, an extra effort is needed to control and manage the field-specific expert knowledge.

2. Companies that have just gone through a major expansion
   In contrast to the start-up phase in which everything was organized informally and on a small scale, the scaling up process in these companies, coupled with a lack of time, often results in employees losing their overview of the activities of colleagues and the organization as a whole. That is why in many such cases the management, but also the employees, get the feeling that they should step back a bit and take stock of the available knowledge. Moreover, all kinds of management processes are often not efficient, leading to frustrations or missed opportunities.

3. Industrial companies that want to enter markets where the required knowledge areas are scattered over the whole organization
   For a separate business unit, the required investments can be considered too high for developing all knowledge areas in-house which are required for competing in markets with complex products. This leads to the necessity of using the expertise of other business units. This is only possible if one can survey the knowledge that is needed and where this knowledge is present.

4. Companies that have completed a re-engineering phase
Following the implementation phase of a BPR-project, many companies often display a drop in effectiveness which is just as drastic as the advantages which were intended when embracing the BPR-philosophy. In particular, experiences from American companies show that this is partly due to poor organization of the processes which must ensure the development, security, and distribution of knowledge within the company. In many cases, the emphasis was only on making the primary processes as lean as possible, but this appears to have an adverse effect on the capacity of organizations to maintain their knowledge household. In retrospect, the middle-manager often turns out to be of great value for managing knowledge!

5. Networks in which several companies and organizations participate
Within a network, several companies and organizations work together to achieve a common goal. A proper combination of knowledge plays a crucial role here, besides a joint investment in the development of new knowledge. Such a cooperation is, however, often frustrated by poor communication between the various parties involved. There is no overview of the available knowledge, there is no common language, there are business and cultural barriers, etc. Explicit and directed control of relevant processes can make an important contribution to improving the knowledge infrastructure of networks.

EXAMPLE: "PROFIT WITH SAFETY"

In Poland, the damage due to industrial accidents is 2–6% of GNP. In order to drastically reduce the total damage from industrial accidents, the Polish Ministry of Social Affairs has initiated the project ARA (Accident Registration and Analysis), carried out by CSC and the department of Safety Sciences.

The ARA project has the following aims:

- Transfer of knowledge relating to the registration, analysis, and learning from near accidents from Netherlands to Poland
- Integration of this knowledge with existing Polish knowledge
- Distribution of the knowledge to Polish companies
- Teaching the companies how to use the new knowledge to learn from the near accidents within their own organization
- Setting up a "knowledge circle" at a national level, within which companies learn from each other and develop new knowledge
- Take measures to distribute this new knowledge to the companies
- Set up a knowledge infrastructure so that Poland can become "self-supporting" in the near future

The organizations participating are companies, trade unions, labor inspection, technological institutes, universities, and the Ministry of Social Affairs. The first phase has recently been successfully completed.

Example: "Body of Knowledge in the Sports World"

In "De telegraaf" the following report appeared: "thanks to an initiative of the NOC*NSF the so-called Body of Knowledge (BOK) was set up last year, in which the available sports science supervision was consolidated, extended and made available to coaches and sportsmen."

"The national coaches of the cyclists lack the necessary knowledge for giving the national teams the ideal preparation required for top-level competition."

APPENDIX 2.1: ACTIVITIES AND PRODUCTS OF KNOWLEDGE MANAGEMENT

Phase conceptual model	Possible activities	Products
Conceptualize "Gaining insight"	• Identify critical knowledge-intensive processes and knowledge bearers and their connection • Describe features of critical knowledge areas and knowledge bearers • Describe interaction of knowledge bearers • Culture analysis aimed at knowledge application and development • Examine communication habits	• Overview of critical knowledge areas • Overview of knowledge per actor • Relation between knowledge intensive work processes • Relation between knowledge bearers
Reflect "Assess qualities and plan improvements"	• Diagnosis of quality knowledge bearers • Analysis of fitness critical knowledge areas • Establish value of knowledge • Simulation of current/desired situation	• Quality model of knowledge in relation to critical demands • Identification of new markets based on existing knowledge • Strong/weak points • Plans for improvement
Act "Actually improve"	• Improve performance of knowledge bearers • Develop new knowledge bearers (e.g., knowledge-based systems) • Develop new knowledge	• New/better interfaces between knowledge bearers • Change of behavior of knowledge bearers • New knowledge bearers • Infrastructure for knowledge sharing • Consolidation of expertise • Better indexing of knowledge
Evaluate "Look back" but also start of a knowledge management cycle	• Measure performance of new or current situation • Determination of bottlenecks and opportunities	• Performance list of knowledge-intensive work processes • Feed back effectiveness and efficiency improvement actions

APPENDIX 2.2: FREQUENTLY ASKED QUESTIONS ABOUT KNOWLEDGE MANAGEMENT

Based on our experiences of the past few years, we have listed the most frequent questions regarding knowledge management, supplemented with a brief reaction.

Is knowledge management the same as information management?

No, information management generally focuses on the final product or knowledge-intensive work processes. It controls processes which are aimed at storing, retrieving, and distributing data.

Knowledge management focuses on the competence of organizations, namely the capacity to interpret data and assign it a value. In addition, knowledge management focuses on another essential product of knowledge-intensive work processes, namely new knowledge. Information management is, however, an important instrument for knowledge management when dealing with the supply of the raw materials for knowledge application and development. Information policy is therefore also a derivative of knowledge policy.

Do theories about learning organizations have the same objectives as knowledge management?

This is indeed the case. KM is, however, a precondition for creating and maintaining a learning organization. Learning organizations sometimes develop spontaneously, for example, in the case of start-up companies. In practice, however, it appears to be much harder to remain a learning organization. Learning is adapting, enlarging, and deepening knowledge, a process that can be controlled through a knowledge management approach. In our opinion, KM is what makes a real learning organization possible.

Is knowledge management the same as knowledge engineering?

Certainly not. Knowledge engineering is a specialist field within information technology. It aims to collect and structure knowledge, programming it into so-called knowledge systems. KE can be an instrument with which the organization of the knowledge household can be improved, but only provides one of the many technologies which can be applied.

Is knowledge management always a lengthy and time-consuming process?

No, knowledge management is very similar to quality assurance. At various ambition levels, improvement actions can be implemented. Using the conceptual model described above it is possible to choose a specific approach which fits the level of ambition.

Is knowledge management the same as "business intelligence"?

No. In general, "business intelligence" refers to the capacity to transform data from the environment into valuable strategic information and knowledge. KM can focus on promoting this "business intelligence" or improving the use of this function, but it is broader than the concept of "business intelligence."

APPENDIX 2.3: ABOUT THE KNOWLEDGE MANAGEMENT NETWORK

The Knowledge Management Network (KMN) aims to promote and support knowledge management within companies and institutions and in a conglomerate of organizations. Improving organizations' business processes and helping them to compete more effectively are our main objectives, starting from an integral and problem-directed approach.

The KMN was founded in 1989. Leading organizations are CSC, Knowledge Center CIBIT, and the Knowledge Research Institute in the U.S. KMN's advisors work in the interface between practice and science and perform projects locally as well as internationally. For further development of our knowledge of KM, we are working closely with leading universities in Europe and the U.S.

REFERENCES

1. Argyris, C. *On Organizational Learning.* Blackwell, 1992.
2. Boer, André de. Knowledge is goud waard, *Personeelswerk*, April 1994.
3. van den Broeck, H. Lerend management; verborgen krachten van managers en organisations, *Lannoo Scriptum,* Tielt, 1994.
4. Danko, Quinty. Concurrentiewapen kennismanagement: Hoe verschillende bedrijven het besturen en beheren van kennis aanpakken, *FEM*, September 1994.
5. Delden, P. J. van. Professionalisering als organisatiestrategie, *M&O*, 1993/3.
6. Drücker, Peter F. *The New Realities*, New York: Harper & Row, 1989.
7. Feigenbaum, Edward A., McCorduck, Pamela. *The Rise of the Expert Company,* New York: Times balsos, 1988.
8. Gardner, Karen. "Position paper for the International Knowledge Management Congress, 1995.
9. Hamel, Gary and Prahalad, C.K. De strijd om de toekomst, *Scriptum Management,* 1994.
10. Koornneef, F., Spijkervet, A.L., and Karczerwski, J. Organisational learning using near-miss and accident data from within and outside your organisation, Proc. of the 4th Safety Critical Systems symposium, Leeds, UK, p. 153–167, 1996.
11. Kratochvil, M. "Developing a know-how strategy" In: J. Liebowitz (Ed.), *Moving Towards Expert Systems Globally in the 21st Century.* Cognizant Communication Corporation, p. 1378–1381, 1994.
12. Laske, O.E. Managerial thinking and knowledge management: a look in the future. WIC Seminar on Knowledge Management, Frankfurt, 1990.
13. Nonaka, Ikujiro. The knowledge creating company, *Harvard Business Review,* 69, November-December, 96–104, 1991.
14. Nouwen, Pieter. Het concern van de toekomst: slank en slim, *Elan,* juli/augustus, 1994.
15. Quinn, James Brian. *Intelligent Enterprise: a Knowledge and Service Based Paradigm for Industry* The Free Press, New York, 1992.
16. Senge, Peter, M. *The Fifth Discipline; the Art and Practice of the Learning Organization.* Doubleday, New York, 1990.

17. Spek, R. van der. Knowledge management: a multi-disciplinary approach to knowledge in organizations, In: *Proceedings IAKE* 1992, p. 225–241. Kensington, MD. Software Engineering Press, 1992.
18. Spek, R. van der and Spijkervet, A.L. Knowledge management; een integrale aanpak voor prestatieverbetering van kennisintensieve werkprocessen Nederland Studie Centrum, 1994.
19. Spijkervet, A.L. and van der Spek, R. Resultaten van een onderzoek naar kennismanagement in 80 bedrijven in Nederland, Rapport Kenismanagement Netwerk, 1994.
20. Spek, R. van der and Hoog, R. *Towards a methodology for knowledge management,* First version is presented on the ISMICK, Compiegne (October 1994). Second version is published as a technical report by the knowledge management network (December, 1994).
21. Schwartz, Peter. *The Art of the Long View,* New York, Doubleday, 1991.
22. Wiig, Karl M. *Managing Knowledge: a Survey of Executive Perspectives.* Arlington, TX: The Wiig Group, 1988.
23. Wiig, Karl M. *Knowledge Management Foundations: Thinking about Thinking. How Ppeople and Organizations Create, Represent and Use Knowledge.* Arlington, Texas, Schema Press, 1993.
24. Wiig, Karl M. A Knowledge Management Framework. Practical Approaches to Managing Knowledge. Arlington, Texas, Schema Press, 1994.
25. Wikström, Solveig and Normann, Richard. *Knowledge and Value,* New York, Routledge, 1994.
26. Zwaan, prof. dr. A. H. van der and Boersma, dr. S. K. Th. Knowledge management, *Bedrijfskunde,* jrg. 65.1993/4.

3 Knowledge-Based Systems: A New Way to Learn

Jay Liebowitz

CONTENTS

INTRODUCTION

In today's environment of corporate downsizing, government restructuring, and defense cutbacks and transitioning, many workers will be in search of a new job or a new career late in life. The organizational hierarchy is being flattened, and many middle and senior managers are being weeded or forced out. Many of these affected individuals are in their late 40s or early 50s, with college tuition and hefty home mortgages to pay. Most of these adult learners do not have the time or patience to enroll in a college degree program to be reeducated or retrained in another discipline.

This is where knowledge-based systems (KBS) fit well, especially in the case of professional transitioning. Knowledge-based systems allow the capturing, encoding, preserving, and distributing of knowledge and lessons learned about a specific domain. If the knowledge bases in these systems were developed well whereby the knowledge is complete, consistent, and validated, individuals could access this knowledge via these systems to get up-to-speed in learning other areas and disciplines. For example, in the case of a worker acting as a defense shipbuilder, as the Navy closes some shipyards these workers will need to be reskilled in other areas. Interactive knowledge-based systems, with intelligent tutoring strategies, may be a feasible way to get this individual transitioned into another area. Or what about the layoff of a defense contractor who is acting as a middle manager or project leader on a defense-related project? Perhaps through KBS, the former managers can access the KBS over the Web and interactively learn another subject area, tangentially related to her discipline. With KBS, the sharing, preservation, and distribution of knowledge becomes possible whereby global knowledge or case bases can be built and maintained for others to use. Xerox, for example, has built case bases (whereby case-based reasoning is used) using 13,000 cases for 90 products to help their representatives worldwide answer questions regarding Xerox products. Reuters has built global case bases whereby financial information is shared and distributed worldwide to its employees and customers. Of course, an important element of developing and implementing these knowledge and case bases is the acquisition and maintenance of the knowledge and cases.

Through KBS development, one can learn how knowledge is acquired and structured. The knowledge acquisition process allows one to refine his/her interviewing skills and organizational abilities in structuring the acquired knowledge. Being able to synthesize the information into a coherent structure helps to enhance one's decision-making abilities by allowing the individual to design, analyze, and evaluate alternatives in problem solving. Additionally, KBS development allows an individual to gain greater insight on how the knowledge of a particular domain is represented. The process of decomposing a domain into its component parts is experienced through KBS development. This decomposition could then expose processes that could be similar to those in related domains which would enable knowledge sharing and reuse to be done. For example, in scheduling domains, there are typically events or activities that need to be scheduled, resources, and constraints related to both resources and events. By developing intelligent scheduling methods to handle the scheduling of these events, resources, and constraints, then some of the representational techniques and scheduling approaches could be generically developed and shared/reused between different intelligent scheduling applications. This ability to share and reuse knowledge is critical to building global knowledge repositories that could be accessed for solving problems or simply for preserving or documenting knowledge before it is lost.

The KBS development process offers a way of thinking about a problem and how best to design and solve it. It is not merely a diagnostic aid or tool, but rather it is a methodology and way of analyzing a task and thinking about its component parts and common processes. KBS development can lead to knowledge sharing and distribution, and when accessed over the information superhighway, or a company's intranet, the knowledge repository can be made available to many others. The Web can serve as the bridges in connecting the islands of knowledge bases.

Let's take an example exploring how KBS can be a paradigm for integrated thinking. Let's assume that we would like to use KBS for improving communications toward building a multimedia software program. Typically, a multimedia project will involve the instructional technologist/designer, multimedia authoring specialist, content specialist, graphic artist, and video/audio specialist (of course, the users and project leader/manager are also included). Each of these individuals may sit around a table and bounce ideas off each other, perhaps even in a distributed mode. If some multimedia projects are limited in budget, all the typical multimedia project team members may not be affordable. If a KBS could be developed which could serve as an interactive substitute for those "omitted" team members, then KBS could serve as part of the integrated teamwork necessary to develop the multimedia program. The KBSs could also serve as multiple knowledge repositories representing the different roles of a multimedia project team, whereby various ideas on multimedia project design could be interactively proposed and managed by the KBSs.

KBS: THE ENHANCER AND SECRET INGREDIENT

Many successful systems use KBS technology, but it is almost secretly hidden. This "raisin in the bread" phenomenon, whereby the KBS flavors the application, occurs often in software development. Many telecommunications fault isolation and diagnosis systems are operationally used, whereby the KBS or artificial intelligent component is 10–20% of the system.

Computer simulation has been used in education, industry, and research as a mode of learning. KBS technology can be used to supplement and augment computer simulations to make them "intelligent." The next section takes a look at ways in which KBS and intelligent systems can be coupled with simulation at the U.S. Army War College.

COUPLING INTELLIGENT SYSTEMS WITH SIMULATION AT THE U.S. ARMY WAR COLLEGE

The Center for Strategic Leadership at the U.S. Army War College has a mission to serve as an education center and high technology laboratory, focused on the decision-making process at the interagency, strategic, and operational levels, in support of the Army War College, Combatant Commanders, and the senior Army leadership. An essential activity as part of the War College student's experience is the use of simulations and models, such as those applied in the 2-week Strategic Crisis Exercise. To date, little work has been done at the War College in coupling knowledge-based and intelligent systems to the simulation and war gaming work. The next few sections will highlight some of the ways that this synergy could be useful at the Army War College.

Simulation

Simulation[1,2] has been used for many years in the military community. Growing out of the operations research and computer science disciplines, simulation has proven to be an extremely powerful technique for modeling, conducting exercises, training, and for other applications. The pilot learning to fly a new fighter jet, the astronaut learning to dock the Space Shuttle with the Space Station, or the negotiation process of military strategic decision making at the various upper levels are examples where simulation can have dramatic impacts. Even using virtual reality[3] technology with simulation is now eagerly being explored.

Although simulation is a powerful technique, there are various limitations. First, most simulations do not have sources of expertise available to the user to aid the user in making improved decisions during the simulation run. Second, most simulations are quantitatively-driven, and may not handle qualitative factors easily. Last, simulations can be computer- and cost-intensive.

Knowledge-based and intelligent systems technology could be married with simulation to address these limitations. Hybrid systems[4] are emerging whereby the coupling of intelligent system technologies to each other and with conventional programming techniques (like modeling and simulation) results. For example, wouldn't it be useful to have an intelligent agent[5] looking over the user's shoulder to act as an advisor for decision making during the simulation. Or wouldn't a knowledge-based or expert system acting as a CINC (Commander-in-Chief), State Department, or the White House staff be useful for tapping their on-line expertise for facilitating the user's ability to make strategic decisions during a simulation. Knowledge-based system technology also handles qualitative modeling and symbolic reasoning well, and can be coupled with quantitative and probabilistic models and simulation. This coupling of intelligent systems and simulation is growing in interest, as evidenced by the panel on "Intelligent Simulation in the Military" at the August 1995 Innovative Applications in Artificial Intelligence Conference.

Although simulation and AI deal with the same issues associated with a specific system, they differ in the way they function: an expert system gives a "prescription" (suggests an action); a simulation model provides a "prediction" (predicts the consequences of a selected course of action under a certain simulation scenario).[10] There is a natural fit and correspondence between AI and simulation concepts, as shown below:[10]

AI	SIMULATION
Object	Entity/resource/transaction/constraint/process activity/event notice/clock
Property	Entity description/data storage
Method	Event behavior
Message	Event execution
Inheritance	Default entity descriptions
Object network	Model/state
Word/context	Scenario/checkpoint
Rules	Constraints/event behaviors/methods to generate experiments
Logic	Constraints/question-answering model/completeness and consistency
Planning	Design of experiments
Diagnosis	Analysis of results
Learning	Detection of causal connections in the model

Nielsen[11] believes that AI techniques can play a variety of roles in the simulation process, including: knowledge representation in a model; decision making within a simulation; rapid prototyping of models; data analysis of simulator-generated outputs; and model modification and maintenance.

Already, the coupling of AI and simulation has been taking place. A coupled knowledge-based/numeric simulation system called NESS (NASA Expert Simulation System) has been developed to help the user run digital simulations of complex dynamic systems, interpret the output data to determine system characteristics, and, if the output does not meet the performance requirements, recommend an appropriate series compensator to be added to the simulation model.[10] Knowledge-based simulation for the military has been conducted at such organizations as Rand Corporation,[12] via the KBSim project. Additional knowledge-based simulation systems are discussed in Reference 1.

At the Center for Strategic Leadership at the U.S. Army War College, the students are usually Colonels who are at the top of their class. The future group of Army Generals will be selected typically from the Army War College graduates. To help these students become better strategic thinkers and integrated decision makers, simulation plays a key role in sharpening their education and skills at the strategic level. The Center for Strategic Leadership (CSL) is the laboratory and high technology center that supports the use of simulation and other technologically-oriented endeavors at the War College and for senior Army leaders in general. CSL has, as one of its groups, an active cadre of operations research analysts who are well-versed in modeling and simulation of military applications. There is also a Knowledge Engineering Group (KEG) that has expertise in artificial intelligence, multimedia, and intelligent systems technologies. Combining the talents of these two groups on projects dealing with intelligent simulation seems to be a natural fit.

The next section will highlight some possible ideas for developing intelligent simulation projects at CSL and the War College.

Possible Collaborative Work at CSL in Intelligent Simulation

There are a number of intelligent simulation projects that could be explored for development at CSL, through the collaborative work of the operations research analysts and the knowledge engineering group. Three such projects will be highlighted next.

Intelligent simulation as part of the strategic crisis exercise

Students at the Army War College are required to participate in a 2-week simulation called the "Strategic Crisis Exercise (SCE)." The SCE is geared to educating and sharpening the strategic decision-making skills through a realistic exercise involving the input from various advisory centers like the White House, National Security Council, CINCs, and other national and multinational groups. The students, using simulation and models, perform role-playing activities in these different "centers."

As part of the computer simulation and role playing, it would be helpful to have on-line knowledge-based or expert systems that could act as knowledge sources relating to a White House Advisor, CINC, National Security Advisor, coalition group, or some other "expert" for assisting in decision making during the student exercise.

Knowledge bases, containing a set of facts and rules of thumb, could be coupled to the simulation to provide the students with "on-line" strategic advice. Multiple knowledge sources could be tied together via the blackboard paradigm to control the agenda and schedule of when a particular source can provide knowledge to the student(s). It is even possible to have multiple, cooperating expert systems where-upon the output from one expert system serves as part of the input of another expert system. By providing the ability to access these strategic knowledge sources, derived from knowledge engineering development efforts with true experts within these agencies, the student should obtain a richer SCE experience and improve their decision-making capabilities.

Another way of encoding these sources of knowledge to be coupled with the simulation is through the intelligent agent approach. People predict that agent-based computing is likely to be the next significant breakthrough in software development.[5] One notion of an agent is a self-contained, concurrently executing software process that encapsulates some state and is able to communicate with other agents via message passing. For those working in artificial intelligence, a stronger notion of an agent is a computer system that, in addition to having the properties identified above, is either conceptualized or implemented using concepts that are more usually applied to humans.[5] In terms of the SCE scenario, the student may have an intelligent agent looking over the student's shoulder to make suggestions and provide advice about the strategic decision in question. As the student runs through the computer-assisted simulation, intelligent agents may be able to provide advice if it looks like the student is "going off in the wrong direction" in terms of trying to solve the strategic problem.

Intelligent simulation, using neural networks/AI, for facilitating
the scheduling of units in a deployed theater

As part of the Crisis Action Model (CAM) and its use in the Strategic Crisis Exercise (SCE), neural networks could perhaps be applied to CAM to help generate the scheduling of units in a deployed theater. Neural network-based scheduling algo-rithms have shown to be used successfully.[12] Just as humans apply knowledge gained from past experience to new problems or situations, a neural network takes previ-ously solved examples to build a system of "neurons" that makes new decisions, classifications, and forecasts.[6] The problems which best lend themselves to a neural network solution are those which do not have precise computational answers but which require "pattern recognition" with "fuzzy variables", as is often the case in dealing with strategic types of problems. A neural network is trained by using historical problem data and once the training process is completed, the network should be able to classify or predict from the new inputs.[6]

Instead of a knowledge-based approach to simulation, a neural network approach could be integrated within the simulation to help, for example, schedule the Army units to be identified and transported to a deployed theater. At CSL, we currently have NeuroShell 2, as the neural network shell. At DuPont, this shell was compared with 11 neural network modeling packages and was the "only neural network shell that met all criteria (associated with a DuPont problem) to make it a useful and efficient tool for the occasional user."[8]

A genetic algorithm or heuristic approach could also be used for this scheduling function, as being explored as part of the GUESS (Generically Used Expert Scheduling System) project.[7] GUESS, which embodies three scheduling methods (Suggestion Tabulator approach where suggestions from the constraints are used for scheduling; Hill Climbing algorithm; and a Genetic Algorithm approach), could be integrated as the back-end scheduler to CAM in order to refine the scheduling of military units in a deployed theater. At CSL, genetic algorithms have already been used and integrated with Excel spreadsheets for advanced course and seminar slating/scheduling. Additionally, through the use of Evolver, a genetic algorithm (GA) Excel spreadsheet add-in, it may be possible to link the other Excel spreadsheet-based models used at CSL with the GA approach for performing a scheduling function.

Knowledge-based simulation for a peacekeeping mission

Computer simulation could be run to help plan a peacekeeping mission. The different officials from the U.S. and the country in question could appear on the computer screen through intelligent multimedia simulations. For example, the student who is learning about strategic peacekeeping missions could run through a simulation where different scenarios operate in real-time addressing specific needs and variations of the peacekeeping mission. Some intelligent multimedia simulations could be developed where the student interacts with the computer, and sees video and hears knowledge from the expert(s) to advise the student of questions that arise during the simulated run. Some initial work in "intellimedia" (integrating intelligent systems with multimedia) has been done by Liebowitz[9] along these lines in building a Protocol Multimedia Expert System to advise students about international protocol (such as the customs, norms, and practices that might affect a peacekeeping mission). The PKI (Peacekeeping Institute) and the KEG at CSL might be interested in such applications.

Intelligent simulation can be a vital strategic area of focus for CSL and the USAWC. Marrying the talents within CSL with the faculty at the USAWC could produce valuable synergistic results. In the coming years, intelligent simulation will play an increasingly important role in not only the military but also in industry and education.

SUMMARY

As we move more into the Knowledge Age, knowledge-based and intelligent systems will continue to play an increasing role at work, play, school, and home. KBS will serve as an integrative mechanism for sharing, communicating, and distributing knowledge across various disciplines. The KBS development methodologies will help individuals acquire, represent, and structure knowledge to facilitate the building of knowledge repositories for permanent usage. KBS will be thought as one of the key technologies in the years ahead!

REFERENCES

1. Fishwick, P. and R. Modjeski (Eds.), *Knowledge-Based Simulation,* Springer-Verlag, New York, 1991.
2. Fishwick, P. (Guest Ed.), Special issue on "Knowledge-Based Simulation," *Expert Systems With Applications: An International Journal* (J. Liebowitz, Ed.), Pergamon Press/Elsevier, Oxford, 1991.
3. Blanchard, D., Virtual reality for business applications, *PC AI,* Knowledge Technology Inc., November/December, 1995.
4. Liebowitz, J. (Ed.), *Hybrid Intelligent System Applications,* Cognizant Communication Corp., Elmsford, New York, 1996.
5. Wooldridge, M. and N.R. Jennings, Intelligent agents: theory and practice, *The Knowledge Engineering Review*, Vol. 10, No. 2, The Cambridge University Press, England, June, 1995.
6. Ward, S., *NeuroShell 2 Description*, Ward Systems Group, Frederick, Maryland, 1995.
7. Liebowitz, J., *Intelligent Scheduling with GUESS for Strategic Decision Support*, U.S. Army War College, Center for Strategic Leadership, Carlisle Barracks, PA, 1995.
8. van Stekelenborg, J., *Evaluation of Neural Network Packages*, White Paper, DuPont Central Research and Development, Modeling and Simulation Group, Delaware, February 6, 1995.
9. Liebowitz, J., S.I. Baek, and D. Finley, "The Protocol Multimedia Expert System," *Hybrid Intelligent System Applications* (J. Liebowitz, Ed.), Cognizant Communication Corp., Elmsford, New York, 1996.
10. Valavanis, K.P. and A.I. Kokkinaki, Knowledge-based (expert) systems in engineering applications: a survey, *J. Intelligent Robotic Syst.,* Vol. 10, Kluwer Academic Publishers, The Netherlands, 1994.
11. Nielsen, N.R., "Application of Artificial Intelligence Techniques to Simulation," *Knowledge-Based Simulation* (P.A. Fishwick and R.B. Modjeski, Eds.), Springer-Verlag, New York, 1991.
12. Cardeira, C. and Z. Mammeri, Performance analysis of a neural network-based scheduling algorithm, *IEEE Spectrum*, IEEE, New York, Vol. 27, No. 6, 1994.

4 Roles of Knowledge-Based Systems in Support of Knowledge Management

Karl M. Wiig

CONTENTS

0-8493-3116-1/97/$0.00+$.50

INTRODUCTION

The basic reason for assembling people in an organization is to give them opportunity to apply their knowledge to perform work to fulfill the organization's objectives. People work to get things done by obtaining information, analyzing challenges, resolving issues, solving problems, making decisions, and implementing changes. Within the organization, work itself is ideally organized to facilitate engagement of knowledgeable people whose expertise is pertinent to the challenges that are faced.* Throughout history, the application of knowledge has exclusively been a human-based function. With the advent of computer-based reasoning and other computer-based functions, this has changed to a considerable degree and our capabilities to automate knowledge and its application are constantly improving.

Partly as a result of our work to deal with inanimate knowledge, we have started to focus on general and specific knowledge assets, their creation, organization, and use or exploitation in ways not previously attempted. We have started to consider how we might manage knowledge in its various manifestations explicitly and systematically. The focus on managing knowledge has also been strengthened by the gradual shift towards a "knowledge society" where in the world-wide competitive market the success and viability of the enterprise — as well as the individual — is based on application of knowledge and the degree of intelligence with which we work.** As a result, to remain in the forefront, enterprises have deliberately started to pursue knowledge management (KM).

KNOWLEDGE-BASED SYSTEMS

Recently, as a result of the systematic perspectives encouraged by explicit KM, the reliance of automated knowledge and reasoning has changed within many organizations. Instead of being considered as stand-alone or relatively isolated solutions to relieve particular critical knowledge-related functions, knowledge-based systems (KBSs) are now often considered as integral building blocks within a larger knowledge management (KM) perspective.

---

* These perspectives reflect views expressed by Terry Winograd (1988).
** Cleveland, Harlan (1985). *The Knowledge Executive: Leadership in an Information Society* and Drucker, Peter F. (1993). *Post-Capitalist Society.*

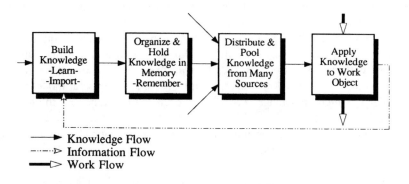

Build Knowledge -Learn- -Import-	Organize & Hold Knowledge in Memory -Remember-	Distribute & Pool Knowledge from Many Sources	Apply Knowledge to Work Object

⟶ Knowledge Flow
┈┈▷ Information Flow
▬▷ Work Flow

FIGURE 4.1.   Four stages of knowledge transition from source to use.

The use of KBS applications for deliberate support of knowledge management is not new. Several practical applications of "expert systems" were fully operational in the early 1980s. Best known among these is Digital Equipment Corporation's XCON system.* Its purpose was to collect, organize, and make available factual knowledge about computer components and expertise on how to configure computers. XCON provided an automatic reasoning framework to assist marketing, sales, and manufacturing functions. The overall motivation and perspective was to make the supported functions act more intelligently by explicitly focusing on managing the selected knowledge according to the four functions indicated above in Figure 4.1. XCON was subsequently expanded to serve additional purposes. Large numbers of KBS applications of many types have been developed and fielded with great success. Many of these are described by Feigenbaum et al. (1988) and Hertz (1988).**

KNOWLEDGE MANAGEMENT — A PERSPECTIVE

Simply stated, the objective of knowledge management (KM) is to build, organize, and make good use of knowledge assets to make the enterprise act as intelligently as possible to secure its viability and overall success. As straightforward as this notion is, to achieve these goals in practice — across all of the enterprise's activity areas — is far from easy. It becomes even more complex when management decides to systematically integrate and manage the important knowledge management-related activities. Each enterprise tends to be unique and options for managing knowledge are legion. Consequently, customized approaches must be devised to provide the enterprise with the best solutions and this adds further to the complexity. Nevertheless, well-established technical options and strategies are available and that alleviates the complexity of pursuing KM. One such important option is automation

---

* Mumford, E. and MacDonald, W.B. (1989). *XSEL's progress, the continuing journey of an expert system.*
** Feigenbaum, Edward A. et al (1988). *The Rise of the Expert Company* and Hertz, David B. (1988). *The Expert Executive: Using AI and Expert Systems for Financial Management, Marketing, Production, and Strategy.*

of knowledge and reasoning in different types of KBS applications for many KM-related purposes.

From a simplified "knowledge life cycle" perspective, the hands-on aspects of systematic KM involve explicit handling of knowledge in four separate functions. They deal with how knowledge makes its way from where it originates (experts, R&D programs, etc.) to where it finally can be used, and a general model for how detailed and explicit knowledge "flows" from each of these functions to the next is indicated in Figure 4.1.

In addition, KM involves several other functions as described below. The scopes of the functions are as follows,

1. Knowledge creation and sourcing — Build knowledge from innovation, learning, and importation. Assemble knowledge from outside and internal experts, R&D, and lessons learned from programs, books, articles, etc.
2. Knowledge compilation and transformation — Organize and hold knowledge in memory and remember. Reconstruct, validate, and inventory the obtained knowledge by organizing it, weeding out outdated and wrong knowledge, etc.
3. Knowledge dissemination — Distribute and pool knowledge from many sources. Disseminate knowledge to where it is needed — to people or embed it in systems.
4. Knowledge application and value realization — Apply knowledge to work objects. Use knowledge to create and deliver products and services.

From a managerial perspective, systematic KM involves four areas of emphasis which focus on top-down monitoring and facilitation of knowledge-related activities, creation and maintenance of knowledge infrastructure, organizing and renewing knowledge assets, and applying (using) the knowledge assets. These areas are shown in Figure 4.2 which also indicates some common knowledge-related practices.

KNOWLEDGE-INTENSIVE WORK

Practitioners and researchers who analyze or redesign how intellectual work is performed are appalled by the general lack of understanding of this area. "How people work is one of the best kept secrets in America" is a statement that expresses this sentiment.* Only in the last few years have we started to understand and focus on the intellectual functions performed by knowledge workers (KWs) when they perform knowledge-intensive work (K-I work). As a result, in many instances we now only start to understand the complexity, power, and business value of how proficient KWs apply the knowledge they possess to analyze and interpret challenges and deliver high quality work products. We are also learning more about how we

---

* Lucy Suchman (1995) in her article on "Making Work Visible."

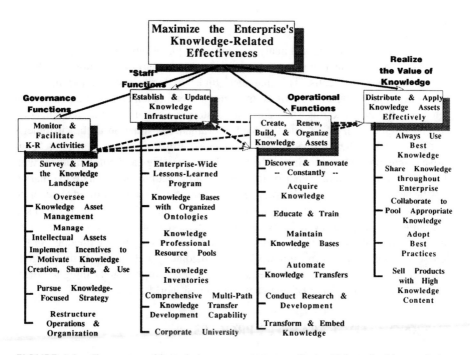

FIGURE 4.2. Four areas of knowledge management emphasis. (Adapted with permission from Wiig, 1995.)

can support KWs to be more versatile by providing them with additional knowledge when needed, reducing the need to educate or train them to become proficient in many rarely encountered or lower-level tasks that can be automated to assist them or be executed autonomously. These are the knowledge-related functions that we often complement with KBS applications.

In most cases, KBS applications support KWs directly in performing K-I work. To do so effectively and successfully requires that the KBS match the KW's cognitive "imagery."* This means that there must be sufficient correspondence between the concepts, associations, mental models, conceptual frameworks, objectives, and driving forces that the KW employs to perform his or her work. The interface between a KBS and its user must be "cognitively efficient" for the KW and must also correspond to the nature of the K-I work as perceived by the KW. From a KM perspective, it must also be designed to support the KW appropriately with regard to the manners in which s/he will use the knowledge, i.e., for planning during a rapidly evolving dialog, for reflective analysis, etc. In addition, the interface must often support other KM functions such as providing feedback to KWs about the results of their work, have flexibility to incorporate new learnings, and support other knowledge building or organizing tasks.

---

* Helander (1990) *Handbook of Human-Computer Interaction.*

An Example

A risk manager in a brokerage analyzes the risk implications and acceptability of a customer's request for placing a large margin trade as part of a complex portfolio with many puts and calls. The manager thinks in terms such as judgments of volatilities in the different securities, matching different elements of the portfolio to determine the best hedging strategies, the customer's collateral and credit history, the credit policies in effect, and the need to evaluate the situation in the customer's best interest. She uses a KBS with the pooled expertise of several experts as "cognitive extension" to her thinking in the investigation strategy and communicates with it by presenting different scenarios that represent her judgment of where the market may be headed. She lets her KBS perform the complex reasoning required to generate the risk evaluations. The KBS reasons in the same manner as she would do, had she had sufficient time and availability to other experts that she would have consulted to complement her own insights. Since the concepts and strategy embedded in the KBS correspond to her own framework, she immediately and intuitively understands its results — no transformation is required — and can immediately integrate them into her exploration. The KBS matches her mental world effectively.

THE KNOWLEDGE TRANSFER PROCESS

Bringing knowledge from the various sources to where it can be utilized — or its value otherwise realized — is a complex process that takes many paths depending upon the nature of the particular knowledge, how it will be applied to deliver products and services, and the preferences or capabilities of the enterprise. If we consider the functions portrayed in Figure 4.2, we see that only the two rightmost functions are associated with dealing directly with the detailed knowledge itself. They constitute the Knowledge Transfer Process. Figure 4.3 shows a simplified view of some knowledge transfer functions and paths. Several end-uses of knowledge are also indicated. Knowledge that is a candidate for transfer may come from innovations, "lessons-learned" programs, research and development, outside sources, or from in-house specialists and experts. Sometimes, such as in apprenticing situations, or when accessing expert networks, particular knowledge may be communicated directly from the source (the expert) to the final recipient (the apprentice). In the situations we examine in this chapter, relevant knowledge is normally obtained from the source, validated, and organized, before it is provided to recipients who may be internal or external users or consumers of knowledge-based products. As frequently observed by knowledge engineers and cognitive scientists, only a limited amount of a person's expertise can be elicited as explicit knowledge for incorporation in KBS applications. The majority of a person's expertise will remain tacit, or automatic knowledge and there are therefore considerable practical limits to how much, and which types of knowledge can be easily transferred between people.* Consequently, we find that critical expertise can only be transferred through apprenticing programs, inclusion

---

* Singley and Anderson (1989) have discussed these difficulties in depth.

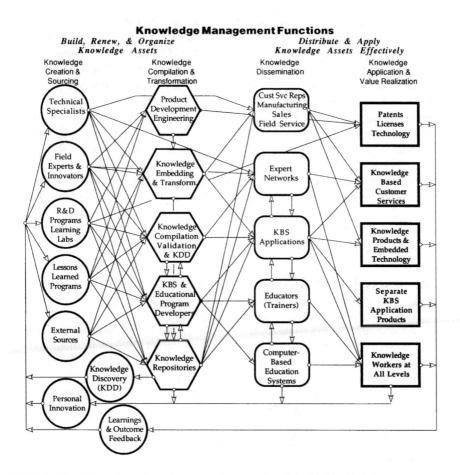

FIGURE 4.3. Examples of knowledge management functions for knowledge sources, transformation, dissemination, and deployment. (Adapted with permission from Wiig, 1993.)

of the expert on the team (through expert networks or in other ways), or transferring the knowledge-intensive tasks to those who already have the expertise.*

KBS applications are able to support several different KM functions in the knowledge transfer process. Most often, they are found to carry expert knowledge directly to KWs for use in a particular work function. However, they are also found to perform other important functions as indicated in the next section.

---

* To facilitate understanding different levels of knowledge, we use four Conceptual Knowledge Levels:
1. Goal Setting or Idealistic; vision and paradigm "care-why" knowledge — "Self-Motivated Creativity"
2. Systematic; system, schema, and reference methodology "know-why" knowledge — "Systems Understanding"
3. Pragmatic; decision-making and factual "know-how" knowledge — "Advanced Skills"
4. Automatic (Tacit); routine working "know-what" knowledge — "Cognitive Knowledge"
From Quinn et al. (1996) and Wiig (1993).

There are numerous dilemmas and tradeoffs associated with knowledge transfers. Some are associated with the large up-front and maintenance costs for acquiring, organizing, and packaging complex knowledge for efficient delivery to large groups of people. Others result from the difficulty of identifying in detail which knowledge is needed and how it will be used to deliver quality products and services. Still others deal with securing and managing technologies for handling knowledge external to people and many of the cultural and social issues that are associated with "replacing people with automation." Additional dilemmas stem from the need to select among the available knowledge organizing methods and transfer mechanisms to find those that the enterprise can handle and be best served by in the long run.

The outcome feedback is a particularly important knowledge flow in the knowledge transfer process. It provides insights into how well products, services, and other work products perform and are received by external and internal customers. Given such feedback, the organization can identify what works well and what does not to be able to experiment and learn how to improve.

SOME KNOWLEDGE-BASED SYSTEM FUNCTIONS FROM A KNOWLEDGE MANAGEMENT PERSPECTIVE

Most organizations that pursue systematic KM in some form also apply KBS applications to support these initiatives. The nature of different KBS applications and the purposes they serve vary widely as do the fundamental roles they fulfill within the KM scope which the enterprise pursues. When we look at the life cycle perspective of systematic KM as indicated earlier, some common functions that KBS are used for include the following examples:

Knowledge Creation and Sourcing — Build Knowledge through Innovation, Learning, and Importation
- Intelligent Outcome Feedback Systems
- Intelligent Agents
- Lessons-Learned Support Systems
- Knowledge Mining Support
- Automatic Knowledge Acquisition with Machine Induction

Knowledge Compilation and Transformation — Reconstruct, Validate, and Inventory Knowledge
- Intelligent Computer-Aided Design
- Knowledge Organization and Restructuring
- KBSs as Repositories for Codified Knowledge
- Intelligent Knowledge Retrieval to Facilitate Reuse of Knowledge
- Provide Capabilities for Knowledge Refining

Knowledge Dissemination — Distribute Knowledge to Where It Is Needed
- Support Education and Training
- Act as Knowledgeable Messengers

Knowledge Application and Value Realization — Use Knowledge to Deliver Products and Services

- Assist Knowledge Workers to Observe, Interpret, Analyze, and Manage Real-Time Situations
- Manage Real-Time Situations Directly with Embedded KBS Applications
- Make Knowledge Available to Assist with In-Depth Problem Solving
- Provide KBS Applications for Sale

Below we will discuss the relationships between these KBS examples and their KM functions in greater detail to elucidate the importance of knowledge automation in the KM context.

KNOWLEDGE CREATION AND SOURCING — BUILD KNOWLEDGE THROUGH INNOVATION, LEARNING, AND IMPORTATION

For the most part, new knowledge is created by people through a number of different processes ranging from ingenious innovation to painstaking and elaborate research. Another process for creation of new knowledge is the uncanny capability people have to see new connections and combine previously known knowledge elements through complex inductive and even abductive reasoning. New knowledge has traditionally been created by people unaided by automation except for performing simplistic tasks such as word processing, filing, mathematical modeling, and statistical computation. This, as illustrated below, is in the process of changing with the arrival of automated reasoning and other knowledge-based system methods. Two areas are of particular importance in this connection. They are the data mining systems and the intelligent agents which both are under rapid development and the importance of which we barely understand at this time.

INTELLIGENT OUTCOME FEEDBACK SYSTEMS

A very important knowledge management function deals with obtaining knowledge about what works — and what does not.

Examples: Fuzzy case-based reasoning (CBR) systems to identify specific classes of intervention strategies that work best (best practices) for different types of mental health and substance abuse patients in different settings.

Automatic and intelligent data mining using limited natural language (NL) processing to locate and reconstruct case histories of successful marketing campaigns from large bodies of marketing data.

Neural net (NN) applications to analyze which control laws provide the best control of a chemical process under different operating conditions.

KM Function: Creates insights not previously available from which greater understanding and knowledge can be derived. Also provides sources of knowledge of great importance for the continued improvement of the enterprise's performance.

Comments: Whereas we yet have not seen numerous KBS applications for creating outcome feedback knowledge, we can expect that this will change when data mining and other technologies become better developed.

INTELLIGENT AGENTS

Examples: Intelligent agents (dæmons) that independently search the enterprise's intranet (or the public Internet) for material on specific topics when provided with appropriate specifications. Many such agents will learn how better to select available material after they bring back their findings to users who then provide feedback on the appropriateness of the selections. Agents may have associated applications to build separate (still unverified and "raw") knowledge bases of selected material for later organization and further selection by the user. Some agents are accompanied by associated applications that will build hierarchies of relations between the collected concepts — at times in the form of formalized ontologies.

KM Function: Intelligent agents normally provide at least two KM functions:
1. Collection of knowledge from a wide variety of sources; and
2. Performing "creative" knowledge selection and aggregation which often result in relations and perspectives not previously available.

Comments: Intelligent agents may become more common as the sophistication of available options increases and additional vendors present their products. In addition, the number of users of intranet and Internet services continues to increase.

We can expect to see average users deploy agents on intranets to locate specific information needed to do their work. More importantly, they may search for — and have the agent organize — knowledge that pertain to their work tasks. Internet users may apply agents to find the least cost airplane connections, good investment opportunities, interesting forums to participate in (there are already several subject areas where the number of relevant forums exceeds a person's capacity to cover them all and therefore will need help to find the ones that provide the best match).

LESSONS-LEARNED SUPPORT SYSTEMS

Examples: KBS applications designed to guide operators in a metal rolling plant who have valuable learnings to recall and view these learnings from operational and corporate perspectives and to describe and document salient features as reference cases to be accessed by CBR applications for later retrieval.

KM Function: Provides an active bridge between the knowledge workers who have noteworthy experiences and insights and the knowledge repository. Supports knowledge elicitation and organization processes and delivers the collected knowledge automatically to the relevant knowledge repository (for example, a "Case Base").

Comments: Lessons-learned support systems provide an important source of new or improved knowledge from knowledge workers' innovations, discoveries, and positive — or negative — experiences. Acquiring lessons learned knowledge is often a major problem. A major issue is associated with the reluctance that people have to depart from their regular tasks to take time to capture the findings. In addition, "writer's block" often impairs the capture process and it must be alleviated as well. The problem can be assuaged when knowledge providers express their insights in the form of "stories" without, or with, a flexible predetermined structure. People often think about and communicate complex concepts and mental models as stories which as a result are powerful formats for relating learnings.*

KNOWLEDGE MINING SUPPORT

Examples: Natural language (NL) understanding programs that examine machine readable text to identify material relevant to specified topics. Intelligent NL systems allow topics to be specified as individual or combinations of concepts and locate text segments which may treat the target concepts explicitly or implicitly.

Case-based reasoning (CBR) systems used to identify patterns of good management practices from large bodies of reference cases that describe the situations, actions taken, and nature and success of the resulting outcomes.

Neural nets (NN) for similar applications of pattern recognition.

KM Function: Provides automatic identification of "best practices" or treatment of relevant topics from historic data such as case histories, reports, or in-house memoranda.

Comments: Many organizations have sizeable libraries of historic reference cases in reports, memoranda, and procedures manuals which can yield valuable knowledge about how specific situations may best be handled. Whereas identification of appropriate best practices may be considered a statistical problem, CBR and other KBS-based techniques provide much more powerful approaches.

NL understanding is still in its infancy, but present techniques are nevertheless quite powerful and some of these are finding their way into commercial uses from having been developed for national security purposes.

---

* Roger Schank (1990) *Tell Me a Story.*

Automatic Knowledge Acquisition with Machine Induction

Examples: The use of machine induction to identify principles of legal rulings in a large body of court cases.

Systems that apply machine induction to search operating histories to identify best practice control laws for operation of physical processes or machinery such as chemical plants, paper machines, or large rotary compressor systems.

KM Function: Provides automatic identification of explicit knowledge from diverse computer-readable sources. The results are often represented in the form of "if-then rules."

Comments: Machine induction has for some time been considered a promising approach to obtain "operational knowledge" directly from operating data and thereby bypassing the "knowledge acquisition bottleneck." However, the method has not yet found widespread use, partly because of its reliance on good and relatively clean operating data.

KNOWLEDGE COMPILATION AND TRANSFORMATION — RECONSTRUCT, VALIDATE, AND INVENTORY KNOWLEDGE

These functions are knowledge-intensive and require considerable expertise to be conducted appropriately. Ordinarily they are performed by skilled practitioners but are now beginning to be supported by "smart" tools that rely on embedded knowledge-based systems. When smart tools are available, they normally are used interactively by knowledgeable professionals who rely on them to perform selected tasks. The benefits of KBS applications for knowledge compilation and transformation can be extensive. However, the benefits are typically indirect (except when there are specific cost savings), qualitative, and their estimation is often based on management perspectives and expectations.

Intelligent Computer-Aided Design

Examples: Support of new product developers with intelligent computer-aided design systems that embed captured and codified (organized using appropriate representations) knowledge of effective technology solutions (such as subassemblies and other building blocks), expert practices, and standards and regulations.

KM Function: Increases the quality of designs by providing effective guidance to designers of when and how to apply known solutions. Also increases effectiveness of obtaining knowledge from various sources and making it available for further transformation to create new products and services for the enterprise.

Comments: Intelligent computer-aided design tools are often designed to provide the practitioners with expert strategies for the overall design effort.

They also may guide designers to create objects that will perform well, can be manufactured easily, and can be produced for a reasonable cost.

KNOWLEDGE ORGANIZATION AND RESTRUCTURING

Examples: Revision and reorganization of knowledge bases using example-guided and inductive logic programming.

KM Function: Transforms knowledge bases from one representation to another to make it possible to use existing knowledge bases for new purposes.

Comments: When we select knowledge from an existing knowledge base for other uses, we often need to transform the content to a different representation. Examples are when knowledge in a lessons-learned knowledge base with structured cases is to provide selected knowledge to a rule-based KBS application or to a computer-based educational program that requires presentation of cases in different formats. In the first case, common "best-practices" rules must be identified from the knowledge base and made explicit. In the second case, general principles and abstractions must be developed from the lessons learned before they are included in the desired format. A second class of applications deals with "truth maintenance," i.e., identifying (and when possible, removal of) conflicts and outdated knowledge in the knowledge base.

KBSs AS REPOSITORIES FOR CODIFIED KNOWLEDGE

Examples: CBR systems and other KBS applications such as rule-based expert systems that build and maintain "corporate memories" and other intelligent databases.

KM Function: Provides memorizing and retrieval functions for relevant and valid knowledge.

Comments: Knowledge repositories need be easily accessible to facilitate user retrieval of knowledge relevant to the situation they work with to observe, interpret, analyze, and handle it. Memorized knowledge must often be retrieved from different perspectives depending upon the situation under consideration and the stage of its treatment and retrieval capabilities are normally designed to reflect such needs.

INTELLIGENT KNOWLEDGE RETRIEVAL TO FACILITATE REUSE OF KNOWLEDGE

Examples: Rule-based KBS in a agricultural machinery manufacturing concern to assist equipment designers identify applicable subassemblies and design solutions for specialized applications.

KM Function: Supports reuse of already captured, codified, and technology-embedded knowledge and its application to create new, customized products.

Comments: A major problem for designers who develop customized products is to ascertain that all pertinent knowledge is reused effectively — and to avoid "clean sheet design" and reinventing. In large companies there are hundreds, at times thousands, of possible "building blocks," each with special characteristics and requirements, creating complex situations that individual designers cannot be expected to know all about. Intelligent design guides have assisted designers in many industries (computer, chemical process design, manufacturing, and even financial services) to find appropriate building blocks and obtain advice on how best to incorporate them in their design solution.

PROVIDE CAPABILITIES FOR KNOWLEDGE REFINING

Examples: Powerful natural language (NL) understanding systems are now available to examine free text and identify underlying concepts of interest for particular purposes.

KM Function: Supports "knowledge refining" by pulling together and organizing knowledge from different sources and perspectives relevant to specific topics and by locating and assembling text and other materials while preserving references to source.

Provides automatic identification and selection of relevant texts such as incoming news items, memoranda, and other documents for automatic routing and prioritization to interested parties.

Comments: Free text documents (books, reports, memoranda, etc.) that treat particular subjects such as news and statistics for crude oil and petroleum trading, contain information which when selectively retrieved, organized, and placed in context constitutes valuable and applicable knowledge. Isolating such knowledge is made highly valuable by being easily accessible and relevant to specific knowledge-intensive tasks.

KNOWLEDGE DISSEMINATION — DISTRIBUTE KNOWLEDGE TO WHERE IT IS NEEDED

Dissemination of knowledge to points-of-action is presently the most important and most frequently encountered function performed by knowledge-based system applications. KBSs provide significant capabilities to support these functions through a number of different approaches. Knowledge dissemination is a well recognized and understood bottleneck in most organizations and frequently prevents both knowledge sharing and use of the best knowledge already available. As a result, the value of intelligent support of these functions is very high — when performed effectively. We have, however, encountered a number of situations where application of KBS technology has not been successful — often because the systems have not addressed the basic needs of the situation at hand.

SUPPORT EDUCATION AND TRAINING

Examples: Interactive multi-media systems with embedded KBS applications to educate sales people and customer service representatives to deal effectively with customers. The intelligent functions rely on feedback from the student to identify progress and identify weaknesses or misunderstandings and present appropriate material to help learners build needed strengths.

Intelligent computer-based systems to educate operators (ranging from airplane pilots to paper-making plant operators) in operating principles for the systems they work with and to train them to understand and manage difficult or rarely encountered situations.

KM Function: Disseminate codified expertise to practitioners who must internalize it for later unaided use in performing their functions.

Comments: Many different KBS applications exist for education and training. Different systems are available for building systematic, pragmatic, and automatic knowledge in preschool children, K–13 students, college students, crafts people, and professionals at all levels.*

ACT AS KNOWLEDGEABLE MESSENGERS

Examples: Model-based KBS applications to accompany complex economic scenarios and impact statements to aid recipients in posing their own scenarios and in interpreting resulting consequences.

---

* Intelligent education and training systems are often designed to monitor learning progress according to the different learning stages (from Wiig, 1993, p. 209).

Stage 1. Perception and comprehension of the knowledge.

Stage 2. Maintenance of the knowledge in working memory until long-term registration occurs.

Stage 3. Registration of the knowledge in long-term store.

Stage 4. Maintenance of the knowledge in long-term store.

Stage 5. Retrieval of knowledge from storage when the information is needed.

Advanced KBS applications will typically support higher level learning modes (modes 3 to 7 below) to help learners develop mental models that provide both general and specific understandings (from Wiig, 1993, p. 209). The seven basic learning modes are as follows:

1. **Rote Learning** or Direct Implanting of Knowledge is an extreme case where the learner accepts the knowledge supplied without examination, judgement, or questioning.

2. **Learning by Instruction** or by Being Told takes place when a teacher provides organized material to the learner who then selects the most relevant facts or transforms the presented knowledge into more useful forms.

3. **Learning by Deduction** is a more complex process where the communicated material contains the subject knowledge implicitly.

4. **Learning by Induction** is a strategy where the learner acquires knowledge by drawing inductive inferences from the supplied material.

5. **Learning by Analogy** makes the learner create new knowledge by modifying specifics of a previously known concept to match the presented material.

6. **Learning from Examples** is when a learner induces a new concept by generalizing from the provided examples, and possibly counterexamples.

7. **Learning by Observation and Discovery** (Unsupervised Learning) is when a learner analyzes observed entities in the provided material and determines that they can be classified into a preexisting or new organizational or representational structure that can characterize or even explain the material.

KM Function: Provides a knowledge dissemination function normally performed by a knowledgeable expert who can support interpretation by an important receiver.

Comments: We often provide expert assistance to assist recipients in interpreting complex analyses when correct understanding and deep insight is highly important. This approach is only feasible when there are very few recipients. When the number of recipients is large, personal assistance is not practical and automated support can be provided.

KNOWLEDGE APPLICATION AND VALUE REALIZATION — USE KNOWLEDGE TO DELIVER PRODUCTS AND SERVICES

Knowledge-based systems are frequently used to directly support realization of knowledge asset values. As indicated in Figure 4.3, end-use of knowledge may occur in many ways and KBS support will therefore vary considerably. KBSs are used to either assist knowledge workers in their work, or used by themselves to apply knowledge to deliver knowledge-intensive work directly and automatically to the workplace. The inclusion of KBS within help desks — ranging from providing customer assistance to diagnosis and trouble shooting — has in the opinion of many become the "killer application" of KBS technology. It is that important.

ASSIST KNOWLEDGE WORKERS OBSERVE, INTERPRET, ANALYZE, AND MANAGE REAL-TIME SITUATIONS

Examples: Case-based reasoning (CBR) systems to support customer representatives interactively with complementary expertise to handle complex, unusual, or unfamiliar situations.

Fuzzy rule-based KBSs to assist insurance underwriters in assessing the risk and evaluate the profit potential for both simple and more complex underwriting situations. (Many other KBS architectures are also used for these purposes).

Structured knowledge base for support of financial services representatives. The KB contains the reference knowledge and is equipped with a special-purpose navigation and knowledge presentation application.

KM Function: Apply knowledge assets indirectly by making them available at the time of use to assist knowledge worker perform knowledge-intensive work without necessarily possessing the applicable knowledge in detail.

Comments: This category of KBS applications is arguably the most important one at this time. As indicated above, help desk applications fall within this category. All types of KBS architectures are applied to assist knowledge workers for a wide range of work situations. Some of these KBS applications are more explicit than others (rule-based systems may contain explicit knowledge while neural nets may contain implicit knowledge), and some are more active (i.e., provide automated reasoning) than others which demand that the user performs all reasoning and only consult the KBS to obtain the "raw" knowledge.

MANAGE REAL-TIME SITUATIONS DIRECTLY WITH EMBEDDED KBS APPLICATIONS

Examples: Rule-based KBS to control autoclave-based chemical process operations and numerous other industrial operations.
Neural net (NN) application to interpret validity and correctness of instrument readings in process control or patient monitoring environments.
NN application to identify letters and words in optical character recognition (OCR) systems for personal computers.
Fuzzy rule-based system to control photographic camera aperture and exposure timing.

KM Function: Apply codified and targeted knowledge assets directly to perform knowledge-intensive work directly and automatically.

Comments: Thousands of applications exist of fully automated reasoning with codified knowledge in numerous application areas, and innovative KBS applications are constantly created in new areas. Presently, application areas range from esoteric uses in unmanned satellites, in all kinds of industries, in business applications of many kinds, in home appliances and automobiles, and in high technology devices of all kinds. It is expected that this application segment will continue to expand for a long time to realize the value of knowledge assets.

MAKE KNOWLEDGE AVAILABLE TO ASSIST WITH IN-DEPTH PROBLEM SOLVING

Examples: Model-based and rule-based KBS application to assist process engineers with preparing input data for, operating, and interpreting results from refinery mathematical programming (LP and MIP) models.

KM Function: Apply knowledge assets to deliver products and services.

Comments: Frequently, in-depth problem solving involves the use of complex methodologies with many steps and intricate, interdependent considerations for specifications of options and interpretation of implications. Practitioners who are experts in the problem domain often are not experts in the details of application of complex analysis tools and therefore benefit from intelligent assistance.

PROVIDE KBS APPLICATIONS FOR SALE

Examples: Expert systems for project planning support created by engineering companies for sale as stand-alone products to architect and engineering (A&E) organizations.

Expert systems for financial planning created by experts within financial institutions for purchase and direct use by consumers for support of deciding how to manage their financial future.

KM Function: Realization of knowledge asset value by packaging the knowledge assets as KBS applications and selling them in the outside market place.

Comments: Most organizations have extensive knowledge assets that they could sell profitably to customers — even to competitors — without hurting their viability or long-term profits. By packaging marketable expertise in KBS applications of suitable architectures, these systems can be profitable additions to the organization's product line.

CONCLUDING VIEWS

Given the broad scope of systematic and integrated knowledge management, it is not surprising that knowledge-based system applications have found many and varied uses in support of KM. What is important, however, is the degree to which KBSs are able to support KM functions. In particular, in well-designed situations and sophisticated organizations, they are very well suited to complement people-based knowledge-intensive activities.

KBS applications for support of knowledge management provide significant benefits. They impact both the quality of results obtained and the effort required to complete the knowledge management tasks. Although the value from the enterprise's perspective are extensive, they may only be known qualitatively since they are not easy to measure quantitatively. The principal value of KBS applications falls into several categories. In general, when KBS applications are appropriate, they perform tasks much quicker and more cost-effectively than most humans are able to do. They are also able to provide deliverables of better and more consistent quality than the average people-performed functions and they can reduce the need for people to be trained to perform many "no-brainer" tasks.

Historically, the major impact of KBS applications in support of KM has been to deliver knowledge to the point-of-action — where the most accurate information on the situation normally is present, analysis is performed, decisions are made, and the opportunity to serve the business in a timely manner is best. The support has either been in support of the KW, or as is the case in many situations, by operating directly on the situation such as when controlling a chemical process. At the present time, we see an increasing number of KBS applications also take on other roles such as to build and organize knowledge, to support education, and many other purposes.

We should be aware of the many and significant limitations of KBS applications. Knowledge management, as most other knowledge-intensive work, relies on innovation and creativity to be effective and provide superior results for the enterprise. Even advanced KBS applications cannot possibly provide these capabilities and at this time are basically destined to perform relatively well-defined functions that by their very nature are more or less routine. Having said that, and as indicated above, we still should be aware that a number of interesting KBS applications are able to create new knowledge and insights by finding and juxtaposing relationships that previously have gone unnoticed.

As pointed out by Dreyfus and Dreyfus, artificial intelligence and KBS applications are no match for the human mind. They examine how powerful proficient performers and experts are in analyzing, planning, and making decisions about complex situations that may contain considerable and unexpected variations. As we have indicated earlier, only a limited amount of a person's expertise can be elicited for incorporation in KBS applications. The majority of the person's knowledge is tacit and cannot be included in the KBS which therefore is limited in its reasoning capabilities compared to the human. However, these limitations are under constant attack by the technical community and KBS applications are becoming smarter all the time.

REFERENCES

1. Cleveland, Harlan, *The Knowledge Executive: Leadership in an Information Society.* New York: Truman Tally Books, E. P. Dutton, 1985.
2. Dreyfus, Hubert L. and Dreyfus, Stuart E., *Mind over Machine: The Power of Human Intuition and Expertise in the Era of the Computer.* New York: The Free Press, 1986.
3. Drucker, Peter F., *Post-Capitalist Society.* New York: Harper Business, 1993.
4. Feigenbaum, Edward A., McCorduck, Pamela, and Nii, H. Penny, *The Rise of the Expert Company.* New York: Times Books, 1988.
5. Hammer, Michael, Reengineering work: don't automate, obliterate. *Harvard Business Rev.* July/August, 104–112, 1990.
6. Helander, Martin, (Editor). *Handbook of Human-Computer Interaction.* Second Edition. Amsterdam: Elsevier Science Publishers, 1990.
7. Hertz, David B., *The Expert Executive: Using AI and Expert Systems for Financial Management, Marketing, Production, and Strategy.* New York: Wiley, 1988.
8. Mumford, E. and MacDonald, W. B. *XSEL's Progress, the Continuing Journey of an Expert System.* New York: Wiley, 1989.
9. Quinn, James B., Anderson, Philip, and Finkelstein, Sydney, Managing professional intellect: making the most of the best. *Harvard Business Rev.* 65, March/April, 71–83, 1996.
10. Schank, Roger C., *Tell Me a Story: A New Look at Real and Artificial Memory.* New York: Charles Scribner's Sons, 1990.
11. Singley, Mark K. and Anderson, John R., *The Transfer of Cognitive Skill.* Cambridge, MA: Harvard University Press, 1989.
12. Suchman, Lucy, Making work visible, *Comm. ACM.* 38, 9, 56–65, 1995.
13. Tecuci, Gheorghe and Kodratoff, Yves, *Machine Learning and Knowledge Acquisition: Integrated Approaches.* New York: Academic Press, 1995.
14. Wiig, Karl M., *Knowledge Management Foundations: Thinking about Thinking — How People and Organizations Create, Represent, and Use Knowledge.* Arlington, TX: Schema Press, 1993.
15. Wiig, Karl M., *Knowledge Management: The Central Management Focus for Intelligent-Acting Organizations.* Arlington, TX: Schema Press, 1994.
16. Wiig, Karl M., *Knowledge Management Methods: Practical Approaches to Managing Knowledge,* Arlington, TX: Schema Press, 1995.
17. Winograd, Terry, *Byte.* 13 (12), December, 256, 1988.

5 Knowledge-Based Systems as a Technology Enabler for Business Process Re-Engineering

W. Joseph Cochran, Arun Vedhanayagam, and Bruce O. Blagg

CONTENTS

0-8493-3116-1/97/$0.00+$.50
© 1997 by CRC Press LLC

INTRODUCTION

Many organizations have been plagued by disappointing returns on their investments in technology and process development. Expenditures on computer technology over the past several years have been enormous, yet productivity improvements have been virtually flat. Information technology has been touted as a strategic competitive weapon, particularly the newer technologies such as knowledge-based systems, but few organizations can point to substantial, tangible advantage. Many approaches to technology deployment continue to be "technology driven," making the tragic mistake of simply overlaying existing work practices with new information systems. Users remain mostly unsatisfied and question the value of information technology.

The scenario for business process re-engineering is similar. Re-engineering has been a disappointment for many organizations, with estimates for failure rates as high as 70%. Even projects that are deemed successful frequently do not reach their projected levels of return. Managing the multidimensional change that is required for successful business process re-engineering has challenged even the most effective leaders.

These are not issues and problems that can be ignored in the near future. Peter Drucker has recently stated that improving the productivity of knowledge workers is a major issue that will challenge organizations for decades to come.[1] We are in the midst of a transition from the industrial age to the knowledge age and those who can harness the power of technology to create knowledge-enabled work environments will lead the way.

Obviously, some organizations have already been successful in unleashing the power of their investments in information technology and have demonstrated dramatic returns in re-engineering initiatives. We have experienced success in both arenas and have learned an important principle: a key to success and to the realization of expected benefits is to **jointly** design and implement process and technology change. In other words, processes must be designed to take full advantage of the current technological environment, and technology solutions must be designed to enable new ways of working. In this chapter, we will describe the role and benefits of using knowledge-based systems as technology enablers in business process re-engineering in some of our recent work.

BUSINESS PROCESS RE-ENGINEERING

Business process re-engineering, as defined by Michael Hammer, is "the fundamental rethinking and radical redesign of business processes to achieve dramatic improvements in critical measures of performance such as cost, quality, service, and speed."[2] Re-engineering is distinct from typical automation projects in that re-engineering challenges the basic assumptions about the current work environment and encourages out-of-the-box, clean slate thinking. It attempts to invent entirely new ways of accomplishing work, not simply minor modifications or enhancements. The objective of re-engineering is to achieve quantum improvements in performance measures, not incremental improvement. However, the most important characteristic of re-engineering is the focus on process, as opposed to a focus on tasks, jobs, or organizational structures.

Processes are a collection of activities that take one or more inputs and create an output that is of value to the *end customer*. The processes that are of interest to re-engineering are core, enterprise level processes, that is, natural, end-to-end processes such as order management, new product development, or manufacturing. Often re-engineering implementations lead to the notion of process orientation, whereby processes become major axis around which organizations are structured and managed.

A typical reengineering project has four phases: (1) defining the process — understanding customer requirements, developing process scope and vision, and setting process goals; (2) understanding the current work — as-is process mapping, analysis and measurement, and benchmarking; (3) new process design — simplifying the process, eliminating non-value added tasks, and developing a technology enabled process vision; and (4) implementing the new process — prototyping, piloting, and rollout.[3] Implementing new information systems can involve similar phases. However, it must be noted that in order for the re-engineering to be successful these phases of process design and implementation must also be complemented by newly designed job definitions, organizational structures, measurement systems, and cultures (i.e., values and beliefs about the work itself).

As can be seen, success in business process re-engineering is dependent upon several complex issues, but one of the most important in our experience has been to pursue re-engineering from the perspective of *technology enabled process change*. Many of the processes of today's organizations were designed at a time when many of the current technologies did not exist. Those processes were probably valid for the environment at that time, but are not adequate now. Processes must be redesigned for and take advantage of the modern technological environment. Knowledge-based technologies are one of the most important mechanisms for enabling a powerful work environment for the knowledge worker.

KNOWLEDGE-BASED SYSTEMS

Knowledge-based systems are computer information systems that explicitly represent and process knowledge. They were the first commercially successful systems to come out of the field of artificial intelligence. The basic insight that led to their development was that much of what human experts know can be encoded symbolically with AI techniques. Experts are distinguished from laymen and general practitioners by their vast task-specific knowledge. This knowledge can be in the form of facts, rules or procedures, heuristics, strategies, and causal domain theories. The science of artificial intelligence provides a means to electronically represent and store each of these forms of knowledge in a *knowledge base*.[4]

Human experts "process" their knowledge with various forms of reasoning: deductive reasoning, inductive reasoning, analogical reasoning, etc. Again, artificial intelligence has given us mechanisms for creating *inference engines* that can process or reason with knowledge. For example, backward chaining techniques can simulate the goal-directed reasoning an expert might use in diagnosis, and forward chaining techniques can simulate the data-driven reasoning needed for a configuration task. In addition, the explicit representation of knowledge processing has enhanced our ability to "standardize" reasoning methods and directly teach mental models regarding them.

Often these subjects, knowledge bases, and inference engines are treated as esoteric by many in the business community. They fail to realize that disciplines such as policy manuals, procedure manuals, and total quality systems are the organization's attempt at manually capturing process knowledge. Procedures documenting "best practices" are used to provide the worker with fundamental rules on how to reason his or her way through the task. Knowledge-based systems provide an interactive real-time alternative to these types of practices.

The capability of knowledge-based systems to represent and process knowledge has moved computers clearly into the realm of knowledge (as distinct from data and information) processing. It has created the ability to support and enable true knowledge workers in the business environment. In other words, we can now automate and support complex tasks that were once considered the unique domain of human knowledge workers. Harmon and Sawyer list five general classes of knowledge-based problems that can be addressed: (1) procedural, (2) diagnostic, (3) monitoring, (4) configuration and design, and (5) planning and scheduling.[5] The examples we will detail later illustrate how knowledge-based systems are used in the areas of diagnostic, planning and scheduling, and configuration and design. To provide a more complete set of general examples, we have broken these general areas into more specific classes and have provided examples of how knowledge-based systems can be used in business applications. (See Table 5.1)

KNOWLEDGE-BASED SYSTEMS AS A TECHNOLOGY ENABLER FOR BUSINESS PROCESSES

Knowledge-based systems have tremendous potential in improving processes. However, they have been mostly neglected by practitioners of total quality management and re-engineering. Knowledge engineers have typically been trained to interview experts and represent their knowledge and current procedures without any attempt to question or challenge the status quo. Therefore, most knowledge-based systems have been designed to support or enhance existing processes rather than enable new ways of performing work. In some cases, this has been acceptable and productivity has been enhanced significantly, but too often it has served to "pave the cow paths."[6]

The value and effectiveness of knowledge-based systems and business processes can be maximized by approaching their design from a joint, integrated perspective. For example, many re-engineered processes utilize a cross-functional team approach. These teams are responsible for executing a variety of tasks that were previously accomplished by separate functions having specialized expertise. This requires the team be made up of "generalists" that understand the entire process and yet know

TABLE 5.1
Classes of Knowledge-Based Systems with Relevant Examples[6]

Classes of knowledge-based systems	Examples
• Diagnosing	• Leading a repair-person through a set of steps • On-line diagnosing of manufacturing equipment • Operating a customer help line
• Planning and scheduling	• Scheduling work flow through a factory • Rescheduling when breakdowns occur • Scheduling events with many conflicting constraints
• Decision-making	• Simulating complex business environments • Rationalizing options for customer promising • Determining the most economical course of action • Analyzing decisions with many conflicting constraints
• Optimizing	• Maximizing factory throughput • Selecting optimal routes for package delivery • Minimizing total cost across multiple facilities
• Designing	• Guiding the steps of a complex design process • Linking CAD software to rules for achieving good design
• Selection	• Selecting among multiple vendors • Selecting product designs • Advising customers on products or options to use
• Monitoring	• Nonstop monitoring of a machine, system, or situation with the application of rules to determine whether an alert is needed for human attention • Monitoring stocks of investments to look for buy or sell indications
• Controlling	• Controlling a chemical plant, blast furnace, or nuclear power station • Alerting human operators or controllers to potential problems on the shop floor
• Interpreting	• Helping in interpreting medical systems • Helping in interpreting market information • Helping to interpret financial/operational performance

how to perform tasks that require specialized expertise. Knowledge-based systems make it possible for the "generalists" in the team setting to have access to expert knowledge and *function* at a higher level of expertise. It also provides this knowledge without the need to stop, contact someone, and wait for a response, which adds speed to the process and increases the productivity of the knowledge worker. This is an excellent way to not only leverage organizational knowledge, but to change a business process and the way people work to improve the value delivered to the customer.

Another hallmark of re-engineered, process-oriented organizations is that it has a much flatter organizational structure. These organizations are typically comprised of specialized knowledge workers who require little supervision (and usually know more than their managers about their job). Knowledge-based systems in this kind of environment ensure necessary structure and standardization in an environment of minimal supervision. Rules can be adjusted and reviewed by steering teams to insure that the proper reasoning is being applied without having to supervise the work directly. Knowledge databases can insure that the process has access to the correct information and is being properly maintained by the knowledge worker.

Another, but perhaps more subtle benefit of the synergy between knowledge-based systems and business process re-engineering is improvement in communication and interactions. Common, easily accessible, structured knowledge bases provide an important foundation for enhanced team interactions. Knowledge becomes a tangible "public" asset to be analyzed, developed, and improved as a vital part of improving a process' ability to deliver better value to the customer. Also, the explicit representation of knowledge affords a common language for more efficient and effective communication. In other words, the use of knowledge-based systems reinforces the mental models and paradigms needed to adjust the culture to the re-engineered process.

Lastly, knowledge-based systems can provide the organization much needed stability when there is personnel turnover. The knowledge worker in today's society is more mobile than any other time in business history. They normally arrive at an organization bringing much needed knowledge which is then augmented with the additional experience received in the process. This makes them more valuable and therefore often more mobile. When a knowledge worker leaves a re-engineered process it often affects the organization's capabilities adversely. This can only be corrected by hiring someone else and waiting until that worker has acquired the experience to duplicate the knowledge of the worker that left. Knowledge-based systems can aid by capturing and preserving much of the essential knowledge and expertise, thereby making the transition for the new worker smoother and improving their expertise faster.

EXAMPLES OF THE UTILIZATION OF KNOWLEDGE-BASED SYSTEMS

The following are "real-life" examples that the authors have personally implemented in several organizations. They show how process performance was dramatically improved using knowledge-based systems. They demonstrate that the effective integration of information technology into the business process can be highly productive and fulfill the long promised advantages of both re-engineering and information technology. As the examples will show, these benefits were achieved when both the technology and the overall process were changed jointly. Changing the technology enablers without major realignment of the process underutilizes the full potential of the new tools. On the other hand, changing processes without significantly improving the technology limits the range of possibilities that can be implemented.

The first example is in the area of planning and scheduling, and demonstrates how a "generalist" team member can be equipped to perform many of the tasks previously reserved for a highly trained specialist. The second example is in the area of diagnostics and demonstrates how having knowledge at the fingertips of the worker can improve skills and reduce overall costs and cycle times. The third and last example is in the area of configuration and design. This example shows how knowledge-based systems can be used to relieve a highly skilled specialist from routine design and project work leaving them free to focus on problems of higher complexity.

EXAMPLE #1: AI-BASED SCHEDULING AT A CAPITAL-INTENSIVE, PROCESS-ORIENTED MANUFACTURING COMPANY

Background

Scheduling is a key business process that many companies need to focus on and optimize. In a high volume manufacturing environment, this process has a significant impact on profitability. The scheduling process is difficult because it must balance the need to meet customer demand "on time" with the need to maximize utilization of resources, minimize costs, and maintain set inventory levels. In this company, scheduling was even more critical due to the fact that the manufacturing process is capacity constrained as opposed to material constrained. Also, the transportation cost is a major portion of the total cost of the product. There are 72 manufacturing lines in 13 locations nationwide, manufacturing 2500 end items which were shipped to over 500 different destinations.

Prior to re-engineering, this activity was carried out by a department named "Planning and Scheduling." Customer demand and forecast were input to this department by the customer service centers. The schedulers developed a schedule based on the demand and forecast and output a production schedule that was then sent to the plants to be executed. The estimated availability was communicated to the service centers who in turn "promised" the customer. This entire cycle consumed 3 to 4 days and was executed daily. Thus, it took 3 to 4 days to give a customer a reliable promise date, inventory levels were high, and the schedules were neither stable nor optimal.

The scheduling software being used was a 20 year old mainframe-based legacy system that was completely manual. It could not be used to automatically assign plants or resources while considering inventory and transportation costs.

Re-engineered Process

Scheduling was re-engineered as part of the order management process. This is the business process that manages the expectation of customers while arbitrating with core business processes such as manufacturing to meet the demand. After re-engineering, scheduling was designed to be carried out by a team called "Tactical Planning." The regional customer service centers were integrated into centralized action teams organized around market segments. These action teams were designed to focus on customer needs and the tactical planning team would arbitrate with other internal processes in order to meet the demands from the customer. All "regular" scheduling decisions, such as assigning plants based on capability, maintaining appropriate inventory levels, capacity utilization, etc., were to be automated. This would allow the action team member to be able to promise to the customer "on-line" while making intelligent scheduling decisions. This would also enable the tactical planning team to concentrate on more difficult issues such as conflict resolution.

Technology Employed

The legacy software hindered the ability to implement the vision of the re-engineered order management process. A knowledge-based finite capacity scheduling software was selected, architected, and implemented in order to be the dynamic scheduler. It was interfaced seamlessly to the sales order processing transactional system (demand), the shop floor control system (production), and the inventory system (inventory). A graphical user interface (GUI) client provided the link between the customer service professional and the various transactional systems and the scheduling system. This was necessary to shield the action teams from the complexity of interacting with different systems simultaneously.

The scheduling system used artificial intelligence based search techniques to resolve conflicts and to optimize the schedule. The search algorithm was based on a unique local repair algorithm.[7] The software is an in-memory application which was ideally suited for dynamic rescheduling since it employed the local repair algorithm. The local repair algorithm minimized perturbation to the topology being operated on by considering the possibility of local changes in order to accommodate a request or change.

When a request was entered by the action teams, the system considered available inventory, available capacity, line capability, transportation cost, and manufacturing cost before scheduling it at the most appropriate location. The total response time from the time of entry of the request to the scheduling decision being made was 2 to 4 seconds. This made the application very suitable for on-line promising of customer delivery dates.

Results

This knowledge-based scheduling software enabled the re-engineered process to implement a new paradigm for scheduling. That is, the action teams were able to schedule the manufacturing capacity on-line considering complex alternatives since the AI software automated much of the decision-making process. Thus on-line finite capacity scheduling was enabled in a dynamic environment. Several manual processes were automated while providing accurate schedules. The action teams were able to interact with each other and with the tactical planning team using a richer more productive environment since the more mundane issues were resolved automatically by the software. The responsiveness to changing customer demands was dramatically improved.

The business results were as follows:

- Turn around time for customer requests were improved from 3 to 4 days to 2 to 5 seconds.
- Effective schedule horizon was increased by 300%.

- The number of people involved in the decision making process was reduced from 9 to 1 or 2.
- Productivity increased over 20%.
- Initial accuracy of placing orders in the correct location was over 95%.

EXAMPLE #2: KNOWLEDGE-BASED SYSTEM FOR REPAIRS AT AN AEROSPACE COMPANY

Background

Repairing aerospace components requires specialized skills and knowledge that is cumbersome to maintain and reference. The representation of the knowledge was not formalized and the learning process was not explicit. Most of the rules and procedures were in the experience of the repair technicians.

Only specialists were able to repair the components. The diagnosis, repair, and inspection processes were inconsistent and based on human memory, thus introducing a high degree of rework and potential problems. These processes were also executed by different individuals, often in different time frames adding to the loss of information and increasing cycle time. Often, there was a long elapsed time between like components arriving for repair, causing repair technicians to go through relearning, increasing the cycle time. The price for the repair was quoted after the analysis process and had to wait for customer approval, adding to the total cycle time.

Re-engineered Process

As part of the re-engineering implementation, teams had been formed and a workflow established to streamline operations from receiving to shipment of parts to be repaired. Technology enablers were implemented in order to facilitate the improvement in turnaround time as well as to formalize the rules and knowledge used in the repair.

The re-engineered process employed one person, enabled by a knowledge-based system, for all phases of the repair, analysis, and inspection processes. The system allowed for the training of generalists that would be able to work like specialists using the knowledge-based system with all the specialized knowledge encoded in it. The process of inspection focused on auditing and enhancing the process of repair and inspection. The price could be pre-quoted since the system standardized the repair process and eliminated unnecessary steps.

Technology Employed

A knowledge-based system was developed that used multimedia to communicate to the repair technician the steps to follow to repair a particular part. A question and answer paradigm was chosen to communicate the knowledge. The system was designed such that it applied generic knowledge on several instances of data thus enabling the addition of new parts without having to change the knowledge base. Explicit learning algorithms were developed and implemented within the system so that learning could progress unobtrusively every time the system was used.

A PC-based knowledge-based package was used as the development environment. This allowed the representation of knowledge as explicit rules and the formalization of typical procedures. Several repair technicians were interviewed in order to "mine" the knowledge from their experience. The rules and procedures were then encoded to form the rule-based expert system.

Data was separated from the rules by employing a standard file naming convention. This enabled the representation of data in standard ASCII files. The rules and procedures were thus true abstractions of knowledge. The rules and procedures operated on multiple instances of data enabling the application of knowledge to various scenarios using the same knowledge base. This also provided the flexibility that was needed in order to add repair procedures for new parts in the same part family. New parts were added simply by adding data files and the system was able to recognize the addition and apply the rules and procedures necessary.

This system implemented a unique learning technique that was unobtrusive yet explicit. The rule weight associated with each rule was modified programatically according to a statistical function which was executed every time a rule was used in the inferencing. This allowed the strengthening of certain rules over others, such that if a certain path during the inferencing was traversed repeatedly then that path was reinforced while others were not. Thus, even though the user was not aware of this, the system was learning over the use of the system. Eventually, the stronger rules fired before the weaker ones, thus speeding up the search. Though no new rules were created, the knowledge base could start with a large collection of rules and yet not compromise on the performance as the system was used over time.

Multimedia components such as scanned images with embedded hot spots, hypertext, and audio playback were used to enrich the interaction with the system. This provided a user-friendly interface enhancing on-the-job training and learning. User acceptance and the encoding of knowledge were faster speeding the development and fine tuning of the system.

Results

The re-engineered process augmented by the implementation of the knowledge-based system resulted in knowledge transfer between skilled technicians. Thus technicians could be cross-trained to be generalists. The learning process was enhanced thus shortening the relearning cycle. The best practices were codified thus improving the learning rate to do better repairs.

The business results were as follows:

- Turnaround times decreased from 2 weeks to under 4 days.
- Rework was reduced to 0.5%.
- Productivity increased 30%.
- Administrative costs were reduced by 30%.
- The cost of quality inspection was reduced by 60%.

EXAMPLE #3: KNOWLEDGE-BASED COMPUTER-AIDED DESIGN TOOLS FOR DEVELOPING NEW PRODUCTS

Background

Developing new products in a high volume consumer packaging industry requires a short design and development cycle. The shorter this cycle, the better the competitive advantage. Rapid development of new and innovative designs ensures participation in a profitable segment of the market. A lack of adequate communication, collaboration, and project management tools led to a noncohesive and nonresponsive new product development (NPD) process. This led to long lead times, long queue times, and unnecessary repetition of work, and it resulted in an environment that was not favorable for concurrent engineering.

Further, the computer-aided design tools were passive repositories of data and were being used more as an electronic drafting board than as an intelligent assistant actively involved in the design process. The knowledge embedded in previous designs were not reused. Designs with minor differences were being designed from scratch every time, leading to long design cycle times and inconsistencies. In order to establish itself as the market leader, the company had to reduce its NPD cycle time drastically and increase the number of new item introductions.

Re-engineered Process

The NPD process was re-engineered to integrate the many departments involved into a more natural, end-to-end process that could be capable of meeting market expectations. The result was a new process that was organized around teams reporting to a single process owner. Though the teams were composed of specialists, technological tools were designed to make the teams cohesive. Customer focus was emphasized and direct customer contact was encouraged. Projects were tracked from the time the need for a new product was identified, through the prototype stage as well as the initial manufacture, until the first batch was successfully used by the customer. Computer-aided tools were used to track, assign, and store the status and cost of projects. Knowledge embedded design tools were developed to actively interact with the designer. Repetitive tasks were automated and part families established in order to speed the design cycle.

Technology Employed

Custom programming under the programming environment in the CAD package was undertaken in order to capture and embed the knowledge that resided among the designers. Parametric design tools were developed so that the designer could work within a virtual design environment that guided the design process. Standard parts of the package were predesigned parametrically so that once key constraining dimensions were specified the component could be instantiated in the design. Once all the components were instantiated, the entire assembly of the container and the mold with its associated components were designed and drafted automatically.

The system guided the designer through the design process by offering a set of choices appropriate at every stage in the design. Once a particular choice was made, it constrained the set of choices available as the design progressed. Thus, the environment presented valid choices to the designer at any time and, by traversing this tree, the designer was able to arrive at a valid design rapidly.

This enabled the representation and sharing of knowledge among the designers. More experienced designers were able to train inexperienced designers "on the job." The designs were standardized and accurate. The cycle time for the design cycle was dramatically reduced (2 days to 10 minutes). The designers were able to focus on issues and designs that required special attention. Quick turnaround on initial design requests enabled the presentation of multiple choices to customers, thus shortening the time it took for customers to decide on any one design.

A Lotus Notes based workflow and project management software package was developed that enabled the sharing of information about all NPD projects. Communications were enabled using Notes Mail within the NPD teams and via Microsoft Mail between the NPD teams and the rest of the organization. A billing system that tracked the number of hours expended on each project and all other associated expenses was developed. Electronic forms in Lotus Notes were developed to communicate relevant information to each team member.

Results

The re-engineered process enabled the collaboration of cross-functional teams during the development of all new products and encouraged concurrent engineering. Cycle time was dramatically reduced due to the timely management of each project. Bottlenecks were identified and adequate remedial actions taken to expedite projects. Knowledge associated with the design process as well as management process was captured and encoded in the CAD system and the Notes system. The employment of knowledge accumulated over the years was facilitated. Standard use of techniques and procedures was enabled. Use of standard components in the design of new packages was enforced.

The business results were as follows:

- The overall cycle time was reduced from 25 to 42 weeks to 9 to 16 weeks.
- The number of new designs was doubled.
- The total number of personnel employed in the process did not increase.
- Successful prototype on the first attempt improved to 95%.
- Successful first production run on the first attempt improved to 99%.
- The billing system enabled the tracking, billing, and collection of over $1,000,000 in legitimate fees during the first year after implementation that had been overlooked previously.

CONCLUSION

In our paper, we have described three examples of re-engineered processes in diverse application areas, order management, product repair, and new product development, in which **knowledge-based systems** were the key enablers for the process. In fact, each of these processes as conceived could not have been fully implemented without knowledge-based systems. In the order management example, on-line dynamic scheduling was a key design criteria for the new process, dictated by the changing expectations of the customer and internal constraints and tension. An AI, knowledge-based approach to scheduling was the **only** way to meet this design criteria and allow the implementation of the new process and organization. In the product repair example, a key process design criteria was the effective and standardized analysis, repair, and inspection of aircraft parts by one person. Knowledge-based systems were developed to be available at each stage of work for each technician, thereby *leveraging knowledge* to enable the generalist approach to the design of this process. In the new product development example, the key design criteria was cycle time improvement and responsiveness to the customer. In this case, knowledge-based design tools were developed to dramatically reduce design cycle time, as well as to improve standardization by making more effective use of the accumulated *design knowledge base*.

We have demonstrated that a paradigm of **technology enabled process change** is a win-win for *technologists* and *re-engineers*. This new paradigm will guide re-engineers to design fundamentally new processes that harness key technologies to create levels of productivity and value that would not be possible otherwise; technologists will avoid the mistake of being technology driven and enjoy the most appropriate, dramatic, and effective use of their tools. The landscape of work will be irreversibly changed and technology utilization will move to a higher plane.

REFERENCES

1. Drucker, Peter F., *Post Capital Society*, Harper Business Press, New York, 1993.
2. Hammer, Michael and Champy, James, *Reengineering the Corporation*, Harper Business Press, New York, 1993.
3. Hammer, Michael, *Reengineering: The Implementation Perspective,* Center for Reengineering Leadership, Cambridge, 1992.
4. Barr and Feigenbaum, *The Handbook of Artificial Intelligence*, Addison-Wesley, Massachusetts, 1989.
5. Harmon, Paul and Sawyer, Brian, *Creating Expert Systems,* John Wiley and Sons, New York, 1990.
6. Martin, James, *Enterprise Engineering,* Savant Press, Lancashire, England, 1994.
7. Zweben, Monte and Fox, Mark, *Intelligent Scheduling*, Morgan Kaufmann Publishing, Inc., 1994.

6 The Role of Knowledge-Based Systems in Serving as the Integrative Mechanism Across Disciplines

Jay Liebowitz, Phillip Kevin Giles, Tom Galvin, and George Hluck

CONTENTS

THE FUTURE IS CLOSER THAN WE THINK

According to Lewis,[1] education in the Information Age is returning to an Agrarian Age form — becoming personalized, individualized, and delivered just-in-time. Educational and developmental paradigms are changing rapidly in moving from old models to new models due to technology implications. For example, lectures are being replaced by individual exploration via networked PCs; passive absorption to projects via skills development and simulations; individual work to team learning via collaborative tools and electronic mail; omniscient teacher to the teacher as a consultant via network access to experts; stable content to fast-changing content via networks and publishing tools; homogeneity to diversity via access tools and methods.[2]

In the past, and still existing to a large extent today, learning is too regimented and compartmentalized. Knowledge is packaged and force-fed into the learning units far in advance of when those units need to know or want to know.[1] It resembles the batched, authority-based assembly line of the factory. For example, a typical college student will take a physics course, chemistry course, biology course, and the like, using a segmented vs. integrated approach. The student learns "batches" of knowledge without usually having a context for applying the knowledge. A useful method would be to have a particular context or issue, like an environmental spill-related problem, and then teach elements of biology, chemistry, and physics as they relate to this issue.

Today's students must learn "how" to learn, and extend learning throughout their lifetime. In today's environment, there is a deluge of data and information, and students must be able to locate, filter, analyze, and synthesize this data and information, and especially focus on the data needed at the moment. According to Lewis,[1] this returns learning to its apprenticeship-based, individualized, just-in-time, Agrarian-style roots. Schank's[3] idea of "active learning, learning by doing" echoes these remarks.

The convergence of technology and its impact on society are the basics of the future. In the next few years, virtual reality and interactive multimedia technologies will be commonplace. Personalized intelligent agents will be available as built-in, on-line experts looking over one's shoulder. Artificial intelligence technologies (expert/knowledge-based systems, speech understanding and natural language understanding user interfaces, sensory perception, and knowledge-based simulation) will be commonly available. Intelligent tutoring systems will be used to allow student-centered, self-paced instruction. Personalized digitized assistants, telecommunications/network advances, groupware, desktop videoconferencing, and collaborative software/group systems technology will be widely prevalent in the next 5 years.

"Virtual residencies" where everyone is connected will continue to emerge for educational purposes. Education will be tailored to individual needs; courses will be available "on demand;" distance learning and "on-line" seminars will be used. The teacher will act as a facilitator, coach, advisor, and collaborator, instead of the sole purveyor of knowledge. Students will have to develop information navigation (searching) skills. And of course, there must be a balance between social and technological interactions.

A key development that is starting to unfold is the replacement of authority-based learning with expert-based learning. Instead of being dictated facts and knowledge through an authority-based learning approach, students will be able to access electronically multiple knowledge bases of expert knowledge at the will of the student. Rather than having one expert speaking to hundreds of people, there will be one computer user with electronic access to hundreds of experts (built-in).[3] The paradigm will be reversed!

Perrin[4] describes the university of the future:

> Imagine a university without walls where you select programs, courses, and mentors from leading institutions of higher learning, libraries, museums, and technical institutes throughout the world. Imagine a university designed for nontraditional learners where learning can take place anytime, anywhere, and where the learner makes the choices. Imagine a university that operates 24 hours a day, 365 days a year where you can participate in live courses or complete the courses in your own time frame. Imagine a university that is truly international, multicultural and multilingual, where courses originate in different countries, cultures, and languages. Imagine a university where computers, interactive multimedia, electronic libraries, and the information superhighway play a major role in providing a full range of interactive courses and services. Imagine a university where the curriculum is oriented to future needs, prepares you for real jobs and initiates placement strategies at the time you enroll. Imagine a university where ingenious, creative, and collaborative efforts are rewarded and programs are future-oriented, exciting, and relevant. Imagine the University of the Future as a virtual learning environment, not a physical campus or place.

This university is not far off in the future. Already many of these ideas are being implemented. An important element of this total concept is the capture, creation, and transfer of knowledge. One of the most critical technologies that can act as the integrating force is "knowledge-based systems (KBS)." The next section explores KBS more fully.

KNOWLEDGE-BASED SYSTEMS: THE INTEGRATOR

As we continue to move into the Knowledge Age, the ability to work "smarter" will be a driving force. A key technology that will help in this regard is "knowledge-based systems" (KBS). KBS are more than just an electronic way of capturing facts and rules of thumb for a particular task or domain. They are not simply systems used for diagnosis, scheduling, classification, or some other functional task. Rather, KBS are a way to help in acquiring, structuring, representing, and encoding knowledge in order to preserve possibly "lost" knowledge and build up the organizational memory of the firm. KBS are a paradigm and methodology that encourages sharing, cooperation, and communication between multidisciplinary elements. In other words, KBS can serve as the integrative mechanism among different disciplines. They and their structure, in turn, can form the bridge among the various islands of knowledge.

Instead of speaking in amorphic terms, let's look at a potential application of using KBS as the integrative element among different disciplines. This proposed example could be used in the classroom.

AN EXAMPLE: LIVING ONBOARD THE SPACE STATION

The problem domain is to help scientists, engineers, astronauts, students, and educators understand how best to live on the Space Station and how to solve certain Space Station-related habitation problems that might occur. This domain lends itself very well to the influence of such multidisciplines as instrumentation and measurement, biotechnology, energy, environmental systems, manufacturing, health sciences, and others. For example, health science considerations could relate to nutritional or medical information that may be needed to solve an individual's health problem onboard the Space Station. Or, certain instruments for measuring environmental conditions onboard the Space Station may indicate potential problems, and the knowledge-based system could suggest ways of remedying these difficulties.

To best implement this KBS, a blackboard architecture could be used. The blackboard model supports the kind of problem solving that is appropriate for complex problems, as in our problem domain of how to live and address problems associated with living on the Space Station. The actual development process in constructing this blackboard-based KBS could follow the knowledge engineering steps: (1) problem selection, (2) knowledge acquisition, (3) knowledge representation, (4) knowledge encoding, and (5) knowledge testing and evaluation.

Problem selection refers to determining whether a problem domain is amenable to knowledge-based systems development and scoping the problem for prototype development. Some of the major criteria for KBS application are (1) the task requires mostly symbolic knowledge, (2) the task takes a few minutes to a few hours to solve, (3) domain experts exist and are willing to work on the project, (4) the task is performed frequently, and (5) the users would welcome the system. In examining the problem domain of diagnosing problems associated with living onboard the Space Station, these criteria seem to apply suggesting KBS appropriateness. Of course, the problem domain as it currently stands is too broad and would have to be scoped into a more manageable subset of knowledge for prototype development. An initial suggestion could be to focus on the problems associated with space manufacturing and related issues onboard the Space Station.

The second step in the knowledge engineering process is "knowledge acquisition." This involves eliciting knowledge from the various domain experts and perhaps others who have direct involvement with the Space Station design and astronautics. Besides interviewing these individuals, various NASA documents, manuals, and handbooks associated with the Space Station effort could be collected and used as part of the knowledge bases (i.e., "knowledge sources"). The knowledge acquisition is the most time consuming process of the knowledge engineering life cycle.

Once knowledge is acquired, the next step involves knowledge representation. Knowledge representation refers to how best to represent the information that the experts and written sources provide. Production rules, frames, semantic networks, cases, and scripts are examples of knowledge representation approaches. A hybrid of these approaches could also be used.

After representing the knowledge, the next critical step is the actual knowledge encoding. As previously mentioned, the blackboard paradigm could be used and an appropriate blackboard shell (e.g., GBB by The Blackboard Group, Cambridge, MA) could be utilized for the prototype's development. In the basic model, a blackboard system is composed of three main components: (1) the blackboard, (2) a set of knowledge sources (KSs), and (3) a control mechanism. The blackboard is a global database (shared by all the KSs) that contains the data and hypotheses (potential partial solutions). The blackboard is structured as a (loose) hierarchy of levels; particular classes of hypotheses are associated with each level, and hypotheses are typically linked to hypotheses on other levels. The levels are themselves structured in terms of a set of dimensions. This makes it possible to provide efficient associative retrieval of hypotheses based on the notion of an "area" of the blackboard. The set of knowledge sources embody the problem-solving knowledge of the system. KSs examine the state of the blackboard and create new hypotheses or modify existing hypotheses when appropriate. The basic control loop is (1) The KS Action component executes a knowledge source instantiation (KSI) (i.e., blackboard events); (2) The Blackboard Monitor identifies triggered KSs (triggered KSs and triggering events); (3) The KS Precondition Components check preconditions of triggered KSs (stimulus and response frame information from successful KS preconditions); (4) Blackboard Monitor updates agenda with KSIs representing activated KSs (updated agenda); (5) Scheduler rates KSIs and selects KSI for execution (selected KSI); KS Action Component executes KSI and the process begins again.[9] A shell like GBB already incorporates the control mechanism and allows for rapid prototyping. In this manner, the development of the KSs can be the main areas of concentration.

The last major step in the knowledge engineering life cycle is "testing and evaluation." Testing could be performed in the following ways: (1) run a myriad of test cases and ask experts to analyze the accuracy of the system; (2) perform a blind verification study (modified Turing test) in comparing the KBS's results with those of experts, with the evaluators being blinded so that they do not know if the results are generated by the KBS or expert. Evaluation could be conducted using criteria such as ease of use, ability to update/maintain, discourse (input/output content), user interface design, natural flow of reasoning, etc. This KBS could also be linked with multimedia to improve the user interface and consultation process.

This example is research that could be performed to emphasize the use of knowledge-based systems as an integrator, via the blackboard approach, across disciplines. One of the leading institutions that has embraced these ideas in developing an "Integrated Science and Technology" curriculum is James Madison University. The integrated role of knowledge-based systems in this new curriculum at James Madison University will be explained in another chapter of this book.

KNOWLEDGE-BASED SYSTEMS: A NEW WAY OF LOOKING AT A SITUATION

Through using KBS methodologies, one should be able to keenly understand a problem or system, model the system components, and comprehend the underlying processes behind the decision-making steps. The knowledge acquisition part of the knowledge engineering process allows the knowledge engineer to dissect a task and understand the relationships between the task's component parts. An appreciation for the steps involved in making a decision and for understanding the influences between these steps is a valuable outcome of the knowledge acquisition process.

An example of the value of such a process is evident in the "Center of Gravity" KBS project developed recently at the U.S. Army War College. A center of gravity refers to those characteristics, capabilities, or localities from which a military force derives its freedom of action, physical strength, or will to fight. A critical part of the strategic level of war is in attacking the enemy's strategic centers of gravity. According to the Doctrine for Joint Operations,[5] there are several purposes to these attacks. They may in themselves be decisive. If they are not, they begin the offensive operation throughout the enemy's depth that can cause paralysis and destroy cohesion.[5]

Determining and understanding the center of gravity concept is both a strategic art and science. In order to bring more structure to this concept, a knowledge acquisition process was used to identify the major steps and their influences in determining the opposing force's center(s) of gravity. From this process, which will be explained in the next section, a KBS was developed.

THE CENTER OF GRAVITY DETERMINATION, ANALYSIS, AND APPLICATION PROCESS[6]

The process for center of gravity determination, analysis, and application is modeled in three basic phases called Situation, Determination and Analysis, and Application. The Situation phase is an analysis of the situation which addresses an assessment of the relevant aspects of the strategic and theater environments. The second phase, Determination and Analysis, describes steps to determine, test, and analyze the strategic center of gravity; and steps to determine, test, and analyze the operational center of gravity. The final phase, Application, describes steps to properly use center of gravity selections to focus war efforts and campaign plans.

The factors and their implications and relationships discussed in the first phase of the model, the Situation, can be very complex and intricate. There may be other factors and other important questions to consider depending on the particular situation. This phase offers a preloading of pertinent information prior to launching into the actual center of gravity determination.

PHASE I: SITUATION

STEP I: CONSIDER RELEVANT ASPECTS OF THE STRATEGIC AND THEATER ENVIRONMENTS

1. Demographic factors
2. Economic factors
3. Geographic factors
4. Historic factors

5. International factors
6. Military factors
7. Political factors
8. Psychosocial factors
9. Interests and political goals

Once the above factors are considered, one must identify all the distinct opposing forces. This is significant because there is one strategic center of gravity per opposing force. Opposing forces are distinct if they are independent with respect to all the above factors. Using WWII as an example, the Axis forces of Germany and Italy constituted one opposing force as they were not politically independent; however, Japan and the Axis forces were completely independent and thus constituted two distinct opposing forces.

Situation analysis concludes with determining the strategic goals and aims of all involved. Goals and aims must remain the main focus during the determination phase (to achieve one's own and deny the enemy's).

PHASE II: DETERMINATION AND ANALYSIS

The second phase consists of four substeps — two relating to the strategic center of gravity and two to the operational center of gravity. The paradigm for both is similar:

- Identify all reasonable center of gravity candidates.
- Test each candidate and select the one center of gravity.

It is important to identify all reasonable candidates first before testing them because the testing process may require several iterations in order to identify a single center of gravity.

IMPORTANT: This phase must be performed for each distinct opposing force identified in Phase I. Remember that each opposing force has its own center of gravity. Perform Phase II on one force to completion before beginning the next force.

STEP IIA: DETERMINE THE STRATEGIC CENTER OF GRAVITY

The strategic center of gravity is most often some controlling aspect of the nation, state, alliance, coalition, or group. One might typically assume that the center of gravity is a political or military entity; however, potential strategic center of gravity candidates can be found within each of the factors listed in Phase I. For example:

- Economic: Centers of commerce or industry.
- Geographic: A country's sole or main port.
- Psychosocial: Will of the people.

There are a number assessments that narrow the list of candidates. Detailed assessments made on the following will help identify reasonable center of gravity candidates very quickly. The order of consideration is important because some early assessments may restrict or eliminate other factors in this list.

1. Composition of Force. Force compositions are classified as alliances or coalitions, single states or groups, or nonallied groups. The composition of a multi state or multi group force may suggest some center of gravity candidates:
    *   Alliances and coalitions are either dominant partner or equal partner, based on whether one force dominates it or if all members share equal power. If dominant, then the combined force likely draws its center of gravity from that of the dominant partner. In this case, only consider center of gravity candidates related to that one force. In either case, the will of the alliance or coalition should be considered a candidate.
    *   Nonallied groups can take many forms, such as rival clans in Somalia; however, their mutual cooperation (or tolerance) is critical if they are to remain a focused opposing force making cooperation among groups a worthy center of gravity candidate.
2. Primary Controlling Element. Ordinarily, the primary controlling element of a force is its governing body (be it a democracy, dictatorship, etc.). In more modern times (especially OOTW), the governing body may be a front for the forces' true source of power, be engaged in a civil war, or simply not exist. Properly identifying the controlling arm of the force is important and will generate numerous potential candidates:
    *   Center of gravity candidates of governing bodies typically include individual political leaders such as a president, king, dictator, etc. It can also include the political cabinet, ruling party, or staff.
    *   Militant groups or clans (not associated with the government military) are likely to have their group leader as the center of gravity. Groups of rival clans may have one dominant clan leader or an alliance.
    *   Illegal economic cartels, such as drug lords, can produce candidates relating to the group leadership or the demand for illegal goods/services. The latter is not likely to be a center of gravity if the cartel is well established because reducing the demand will probably not affect the cartel's ability to function.
    *   Legal businesses or groups would probably be the primary controlling element if their economic impact on the force is so strong that the governing body is in no position to oversee or control them. In this case, the center of gravity candidates could include the CEO/Board of Directors, Stockholders or Stakeholders in the business, or demand for goods/services.
3. Type of Government. The will of the government is a strategic center of gravity candidate. Assessing the type of government of an opposing force will provide strategic center of gravity candidates. There are three basic types of governments:

- Democracies. Democracies can be representative, such as the U.S., or parliamentarian, such as Britain. The will of the people is a candidate for both, and the will of the parliament is a candidate for some parliamentarian democracies.
- Totalitarianisms. These can be either military dictatorships or police states. For dictatorships, consider the military element, dictator, and/or staff as candidates. For police states, consider the police element, political leader, or staff.
- Feudal Societies. Feudal societies are headed by a god/king, which is the obvious center of gravity candidate for these types of governments.

4. Level of Civilization. Levels of civilization help identify both the economic growth and prosperity of the nation/state and identify the forces' ability to sustain itself during a conflict. There are three levels:
   - Pre-Industrial, also known as agrarian. Pre-Industrial societies rely heavily on the strength of its Capital, which makes it a strong candidate.
   - Industrial. Industrial societies are much more diverse; however, they may still rely on a particular center of commerce or center of industry. Centers of commerce could include a main or sole seaport or airport. Centers of industry might include a main hub for the manufacture of some strategically important goods (like nuclear fuel or warheads).
   - Informational. Because the "informational" or media-driven society is still a new concept, those nation/states that can be considered "informational" are still likely held to their industrial roots. Therefore, in addition to considering the societies' information networks or systems, also consider the industrial candidates.

5. Other Factors. Additional strategic center of gravity candidates can be derived through analyses of other relevant factors. These candidates include a special strategic capability such as a nuclear threat; key nonpolitical or independent figures such as religious leaders, orators, activists, or special interest organization leaders; and others.

The above analysis should lead to a menu of potential strategic candidates. The test for a strategic center of gravity must then be applied to each candidate:

> Can imposing your will (destroy, defeat, delay) on the potential center of gravity candidate create the deteriorating effect that prevents your foe from achieving his aims and allows the achievement of ours ... and will it be decisive?

The single center of gravity candidate passing the test is selected as the strategic center of gravity. A frequently asked question at this point is "what if more than one center of gravity candidate passes the test?" The authors' interpretation of Clausewitz and their belief in the existence of only one strategic center of gravity suggest that more than one candidate cannot pass the test. The test has been incorrectly applied if a particular assessment indicates more than one strategic center of gravity candidate passing the test. Granted, this is a contentious issue.

Step IIB: Analyze the Strategic Center of Gravity

This critical step addresses need, ability, adverse effects, and willingness of applying direct military action or other approaches to influence the center of gravity. Included in this analysis are the following questions:

- Is the engagement total war or something less? Direct military action is necessary if the engagement is total war.
- If engaged in something less than total war, is it essential to destroy/neutralize all of the center of gravity? Direct military action should be considered the strongest option if all of the center of gravity must be destroyed. If not, direct military action may not be essential and other approaches should be considered.
- Does one have the ability to directly impose its will on the selected center of gravity?
- Will a direct attack cause adverse second and third order effects?
- Is the political leadership willing to directly and decisively engage?

Failure in any of the last three questions above indicates the need to either reassess ends/means and interests/objectives, or reassess strategic goals and aims before committing to direct military action (to preclude the repeat of historical mistakes). Once various approaches to influence the strategic center of gravity (including direct military action) are deemed viable, major operations and goals are identified.

Step IIC: Determine the Operational Center of Gravity

This step begins with determining operational goals and aims of the opposing force to insure proper focus is maintained during the analysis (to achieve our goals and deny the enemy's). Identifying the operational center of gravity is much simpler than the strategic counterpart. An operational center of gravity is most often described as some dominant characteristic of the opposing force. Operational center of gravity candidates include, among others:

- A dominant allied force
- The entire armed forces of an allied nation or group
- A dominant joint service force or capability
- A dominant service or capability
- A dominant element within a service
- A dominant capability of a service element
- The threat of intervention from a new power

Each operational center of gravity candidate must be tested against the following: "Will successful action against the selected candidate decisively achieve our aims and deny the enemy's? And is it the most focused choice?"

The candidate passing the test is selected as the operational center of gravity. Properly applying the stated test will result in only one most focused operational center of gravity.

Step IID: Analyze the Operational Center of Gravity

This step begins with determining ways to influence the operational center of gravity and the decisiveness of immediate action. This assessment (similar to that conducted during the strategic center of gravity analysis) addresses the ability, will, and need to immediately act and impose one's will on the operational center of gravity and inputs the operational center of gravity selection into the operation planning process. Courses of actions that maximize indirect influence over the operational center of gravity should be favored if one cannot immediately act and impose its will on the operational center of gravity. If one can immediately act and impose its will on the operational center of gravity, favor courses of actions that provide direct influence.

Both cases involve assessments of related decisive points and key vulnerabilities. Remember that decisive points and key vulnerabilities are not centers of gravity! Staff estimates are conducted after the decisive points and key vulnerabilities are determined.

PHASE III: APPLICATION

Before continuing, the friendly center of gravity must also be known. In the Application Phase, the plan must not only include the methods by which the friendly forces will impact the center of gravity, but also the way the friendly forces will protect its own center of gravity from enemy influence. The process to determine the friendly center of gravity is the same as that used to determine an opposing force's center of gravity. If needed, perform Phase II using the enemy's perspective to determine the friendly center of gravity.

Once the centers of gravity for all opposing forces have been determined and analyzed, they are used to focus war efforts and campaign plans. Once the plans have begun execution, the situation must continuously be reassessed to detect potential changes or shifts in the enemy center of gravity. This reevaluation loop consists of the following items:

- New elements enter the conflict. The situation must be reassessed and the centers of gravity redetermined if there are any new elements (allies, coalitions, members, or groups) who have or might enter the conflict. If so, return to Phase I.
- There are changes or shifts in the campaign plan. The operational center of gravity for the opposing force may need to be reassessed based on campaign shifts or phases. In other words, the operational center of gravity will probably be different for each phase or campaign because goals and objectives will more than likely be different. If this is the case, return to Step IIc of Phase II for the opposing force in question.

- Changes in capabilities or aims. The operational center of gravity needs to be reassessed based on new forces entering the theater, significant technology changes (i.e., weapons of mass destruction), or the evolution of new aims. If this is the case, return to Step IIc of Phase II for the opposing force in question.

A frequently heard argument asserts that gaining new information is another situation that may cause the operational center of gravity to change. This argument is invalid because unique centers of gravity exist independent of one's information or knowledge about them. The original assessment and determination led to the selection of an incorrect center of gravity if new information indicates the existence of a different center of gravity.

Certainly if any item changes from Phases I or II that link directly to the selected strategic or operational center of gravity, the continued validity of center of gravity choice should be confirmed.

SUMMARY

Knowledge-based systems, and their knowledge engineering development process,[7,8] are extremely useful for capturing, sharing, and preserving knowledge. A new way of thinking needs to be considered in viewing KBS as not just diagnostic aids, but rather as a way of better understanding the systemic processes involved in a particular task or domain. KBS could serve as the integrative element among different disciplines to facilitate the communication, distribution, and sharing of knowledge, as described in the "Living Onboard the Space Station" example. The knowledge acquisition process facilitates a deeper understanding of the functional steps and considerations involved in a given task, as shown in the Center of Gravity example.

As we look toward the future in the Knowledge Age, KBS technology will play a paramount role. Very few decisions can be made using isolated pockets of knowledge. An integration among these different knowledge sources is, and will be, needed. This is where KBS will have the greatest impact!

REFERENCES

1. Lewis, T., Living in real time, side B, *IEEE Computer,* Institute of Electrical and Electronic Engineers Computer Society Press, Los Alamitos, CA, October 1995.
2. Reinhardt, A., "New Ways to Learn," *BYTE*, McGraw-Hill, New Hampshire, March 1995.
3. Schank, R., Keynote address on "Active Learning", International Symposium on Artificial Intelligence, Monterrey Institute of Technology, Mexico, October 1995.
4. Perrin, D., The university of the future, *Educ. J.,* Vol. 9, No. 2, February 1995.
5. U.S. Joint Chiefs of Staff, Doctrine for Joint Operations, Joint Pub 3-0, Washington, D.C., February 1, 1995.
6. Giles, P.K. and T.P. Galvin, "Center of Gravity: Determination, Analysis, and Application," U.S. Army War College, Center for Strategic Leadership Publication, Carlisle Barracks, PA, 1995.

7. Liebowitz, J. (Ed.), *Hybrid Intelligent System Applications,* Cognizant Communication Corporation, Elmsford, New York, 1996.
8. Liebowitz, J. et al., *The Explosion of Intelligent Systems in the Year 2000,* International Society for Intelligent Systems (ISIS), PO Box 1656, Rockville, Maryland 20849, 1996.
9. Vinze, A. and A. Sen (Eds.), Special issue on "The Blackboard Paradigm and its Applications," *Expert Systems With Applications: An International Journal* (J. Liebowitz, Ed.), Vol. 7, No. 1, Elsevier/Pergamon Press, New York, January-March 1994.

7 KBS in Integrating Engineering and Construction Projects

William J. Barnett

CONTENTS

INTRODUCTION

The engineering and construction of a major industrial plant, such as an oil refinery, power plant, or automobile assembly plant, is a large, complex, and costly project. These endeavors generally cost in the 100 millions of dollars and can reach several billions of dollars. The number of engineering personnel is typically many hundreds, and the total construction and project management personnel reach into the many thousands. The number of different pieces of equipment, miles of cable, tons of steel, feet of pipe, doorknobs, bolts, etc., is staggering. And yet, each one of these items must be designed, specified, procured, delivered, constructed/installed, and made to work. The plant shown in Figure 7.1 illustrates the size and complexity. Note the numerous vessels, tanks, structures, etc.

Computer technology and telecommunications systems are critical elements in making these projects happen. Computer-aided design (CAD) systems and database systems, along with local area and wide area networks, are the key technologies in use today. But, knowledge-based systems (KBS) that can automate or semi-automate much of the work are beginning to be introduced, and will eventually have a major impact on the engineering and construction (E&C) industry.

0-8493-3116-1/97/$0.00+$.50
© 1997 by CRC Press LLC

FIGURE 7.1.   Picture of a large new facility (e.g., oil refinery).

Fluor Daniel, which is one of the largest engineering and construction companies, has for several years, been developing a wide variety of database, CAD, KBS, and other technologies that can reduce costs, improve quality, and speed up the activities on projects. During this time, we have evaluated most of the new technologies that are currently available and have seen how a wide range of other companies are using these technologies. We have acquired those that worked for us, and have developed, sometimes painfully, many applications and systems. We are actively using many of these new technologies and applications, but we are not finished. We probably will never be finished because new concepts, technologies, and problems seem to occur faster than the solutions.

TYPICAL ENGINEERING AND CONSTRUCTION PROJECT

As stated earlier, the design and construction of a large industrial project is a very complex undertaking. The project team to perform these projects is typically organized as shown in Figure 7.2. Note the large number of different disciplines and functions that are required, and then imagine all of the coordination needed to keep all these groups working properly. Each of these disciplines and functions perform different activities and produce different information. However, each of these groups use information, data, and decisions produced by the others. The timing of the generation of this information must be orchestrated very carefully. It is the management of the timely production of this information and decisions based on the information that is the critical role of project management.

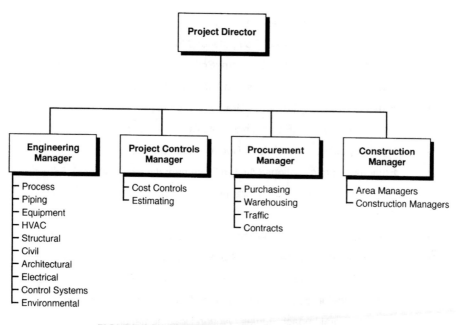

FIGURE 7.2. Organization chart of a project team.

An important factor in managing a project is achieving a high degree of integration of the people, organization, and work processes that facilitates the production of the needed information. At one time, this integration was achieved by having all the project personnel located near one another in a "task force" area. By being physically close, communications among the participants is much easier and faster. With the advent of high speed networks, along with e-mail, databases, etc., good communications and the integration of information on a project can be achieved without having everyone in one location. At the current time, the task force concept is often used, but there is a trend toward having project teams located in multiple areas, buildings, and even countries. There are good examples of projects where the design work is performed in several different countries sharing data, information, and drawings across time zones.

There are times when these well-managed teams are challenged, especially when major changes are made that impact much of the design. The handling of changes has a ripple effect that can upset the best of design teams, and can challenge the best of the design computer systems, because all of the data that is impacted must be identified, and all the decisions that effect other decisions must be made properly. One of the major advantages of three-dimensional computer-aided design (3-D CAD) systems is the ability to detect the interference of one element into another. Unfortunately, there are many types of errors and omissions that systems cannot detect. Knowledge Based Systems, integrated into project databases, offer advantages in helping solve this problem.

There are three significant information systems and work process parameters: (1) the many different functions performed by the many people on the project, (2)

- Process Flow Diagrams (Schematic)
- Process Model (Chemical Computation Model)
- Piping & Instrumentation Diagram (Schematic)
- Equipment Layout (Drawings and Graphical Models)
- Structure Design (Drawing & Graphical Models)
- Structural Analysis (Engineering Calculations & Model)
- Piping Specifications (Written Documents & Database)
- Equipment Specifications (Written Documentation & Database)
- Piping Details (Drawings & Graphical Models)
- Electrical One Line Diagram (Schematic)
- Electrical Short Circuit Analysis (Engineering Computation & Model)
- Wiring Diagrams (Drawings & Graphical Model)
- Wiring Connection Lists (Written Documentation & Database)
- Instrumentation Specifications (Written Documentation & Database)
- Environmental Impact Studies (Written Documentation)
- Site Plan (Drawing & Graphical Model)
- Architectural Concept Studies (Drawings, Graphics)
- Control Systems Functional Requirements (Written Documents & Database)
- Control System Sequence of Operation
- Project Cost Estimates
- Cost Report
- CPM Schedule
- Progress Reports

FIGURE 7.3. List of deliverables on a project.

the variety in the types of data and information involved in this work, and (3) the enormous volume of information generated and used throughout the projects. As an indicator of these factors, see Figure 7.3, Project Deliverables, for a partial list of the documents or databases produced during the project. These deliverables range from 3-D CAD models, to traditional drawings, to pages upon pages of specifications, to cost reports and purchase orders, and include mountains of meeting minutes and reports.

This data and information must be communicated and made available to team members, suppliers, subcontractors, and clients located around the globe. The large quantity of data makes it imperative to have good computer and communication systems.

For KBS systems to be effective, they must operate in these environments, and hopefully provide very useful functions that make people want to use them.

PROJECT COMPUTER SYSTEMS

A large E&C project lives on data, information, and knowledge. Although the final product is a newly constructed facility of concrete and steel, a very large portion of the work is information processing. Even the construction craftsmen, who are the people that convert the information into a physical plant, need lots of information and data. The engineering, procurement, project controls, and project management functions are almost totally involved in information and data. Everything they produce shows up on some document, database, or computer model.

With this heavy requirement for information, computers have to play a major role. Indeed, Fluor Daniel and the entire E&C industry were very early users of computer technology. For many years, mainframe computers were used for engineering calculations, scheduling, cost accounting, and even some of the early CAD systems. As computer technology progressed, the usage within the E&C industry increased rapidly. The introduction and steady improvement of PCs along with low cost networks, have caused a veritable explosion of new applications. In the future, a large portion of those systems will be Knowledge Based Systems (KBS), but at the current time most are the more traditional applications.

A small sample of these are shown below:

Project deliverables	Systems	General applications
Process Modeling	Workstation	Scientific Calculations
2-D CAD Drawings	High-End PC	Graphics
3-D CAD Model	Workstation	3-D Graphics
Electrical Short Circuit Analysis	PC	Engineering Calculations
Structural Design	High-End PC	Finite Element Analysis
Piping Specifications	PC	Word Processing/Database
Wiring Schedule	PC	Word Processing/Database
Flow Meter Orifice Calculation	PC	Engineering Calculations
Project Schedule	PC	Critical Path Method (CPM)
Cost Estimate	PC	Spreadsheet/Database
Cost Accounting	PC and Mainframe	Accounting System
Purchasing Contracts	PC	Word Processing
Procurement Activities	PC	Database

In most instances, all of these applications need information and data from the other applications, so the systems are networked together. And, since the construction site, the owner, and the various other suppliers are at other locations, this networking is very important. A schematic for a typical project is shown in Figure 7.4. This shows just one "home" office, but there may be several offices working on a single project. The computer system on a large E&C project can be a complex network of advanced machines stretching around the globe.

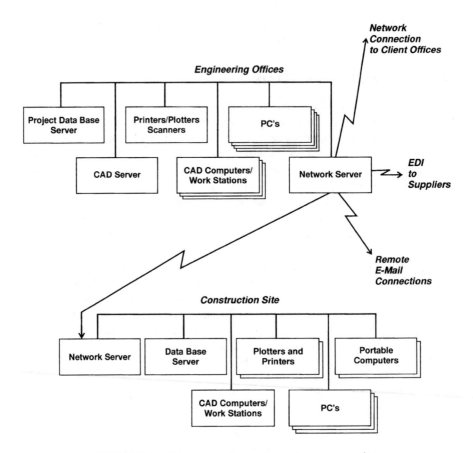

FIGURE 7.4. Schematic of computer systems on a project.

The biggest concentration of systems will be in the engineering home office(s) where practically everyone working on a project has a computer. Most of these will be a general PC class that runs the typical office suite of programs, plus a range of other engineering calculations and database applications. Some applications will require more powerful systems, especially 3-D CAD systems and process modeling systems. Other areas that require extra hardware are the large database and network servers and accounting systems.

As discussed earlier, the overall computer system is an important element in the integration of the work process, and the results of those processions into the project deliverables. Good computer systems can significantly improve the performance of an E&C project team. Personnel can have fast and easy access to data, can produce designs using computer tools, and can communicate results to the entire team. However, computer systems are costly, and the E&C industry has had an arduous task in developing and applying systems that perform better and reduce the cost of work for their clients while also making money for themselves.

BUSINESS ISSUES

The tremendous growth of computer technology in the E&C business is driven by a number of factors. Sometimes, the availability of new and interesting technology and the hype produced by the computer industry causes people to buy new products without a good reason. But, in the long run, large investments in new technologies must make business sense. They must make the purchasers more competitive and more profitable.

Within E&C firms, there are conflicting priorities. While we need to have greatly enhanced capabilities, we also need to keep our costs down. Engineering and other home office labor is one of the major expenses related to a major new facility, and owners of the new facilities are continuously looking for ways to reduce those costs. Therefore, E&C companies must reduce the labor costs required to perform the work. Computer systems, and design automation applications should help reduce these costs, but the costs of their technology must also be considered.

Technology is expensive. In their day, mainframe computer costs seemed far too high, so that when personal computers (PCs) were introduced, it appeared that computer costs could be significantly reduced. However, as the number of PCs have grown, and the cost of networks, support, and maintenance has increased, the overall costs have actually increased. Add to this the rapid technological obsolescence of PC hardware and software, with the need for continual training, and it is clear that cost containment has to be a major emphasis.

The reason for this discussion is that knowledge-based systems represent another major development and support costs. And, while the potential payback is very large, the costs of developing a comprehensive KBS is also large.

THE GRAND VISION

The E&C industry is an information-intensive business, and at the current time, computer and communication technologies are widely used. However, much more can be done. Much of the engineering and home office work can be made more efficient by using traditional computer technology. And, many of the activities associated with engineering analysis and design can be improved and automated with KBS systems.[1]

Several years ago, Fluor Daniel established an organization and a program to evaluate and develop new technologies and methods. This group was interested in any ways that could significantly reduce costs, shorten schedules, and improve the quality of projects. Much of the emphasis was on improving the work processes prior to attempting to automate them. In addition, we looked at methods for automating construction and other related concepts. However, the largest part of the program was focused on developing or acquiring advanced computer systems.

The basic concept was to try to use computers and KBS systems to the greatest degree possible to perform those activities which could be captured as a knowledge base or a database. This would then let the engineers and other professionals concentrate on performing the knowledge intensive and innovative activities. We knew that the concepts had been proven for smaller size activities, but we did not know if the technology could handle the complexity and size of the projects we deal with. Nor did we know how much all of this would cost or how long it would take.

As a part of the study, an analysis was made of the work processes of all the disciplines and functions involved in a project. In addition to the knowledge gained about the detailed work activities, there were a number of themes that emerged.

1. Much of the engineering work is based on either, (1) well-known scientific principals, or (2) published codes and practices. A large part of the remaining detail work is based on documented best practices. However, there are still many decisions that are based on undocumented methods and personal preferences, and there are sometimes several conflicting "best practices."

2. Much of the work is similar to previous projects, both as an overall project, and more so, in the details of the various discipline groups. Therefore, much savings in work can be achieved by quickly accessing examples of work from previous projects. (However, all engineers know how dangerous it is to consider a project as a copy of a previous project, because a few small changes in the final product can result in numerous changes in the details.)

3. There is a significant amount of creativity, innovation, and judgment involved in the design process that is not documented and will be difficult to capture as "knowledge." The amount of this type of knowledge is very difficult to measure but the author estimates that around 10 to 20% of the overall work activities are in these categories.

4. Significant benefits can be received by having fast and easy access to all the current data and information. This would reduce delays in obtaining information, and would improve quality by always having the most recent data.

All of these trends provided encouragement that KBS technology could significantly reduce the costs of engineering and home office labor. It looked like all we had to do was to develop a series of KBS applications along with several new databases and 3-D CAD applications, and we could be well on the way to a much improved project delivery method. These new systems would reduce the costs of engineering, cut the schedule to fractions of the current time, and provide better information to the construction project.

The set of technologies that made up this "grand vision" included the following:

1. Knowledge-based systems to automate the many repetitive, noncreative activities. The types of knowledge representation included rule-based expert systems, semantic nets, frame, and object-oriented programming technologies.

2. 3-D CAD models, attached databases, and KBS systems that can model every significant component of the plant.

3. On-line database systems for storing all data, for rapid access by any authorized personnel, and for managing changes throughout the life of the project.
4. Document management systems that allowed fast and easy access to data and documents of previous projects.
5. Process and operations modeling and simulation systems that could accurately predict how the new plant would operate.
6. High speed data communication systems, portable computers, etc. that allowed everyone who needed it access with the databases and the ability to work remotely with others.

There are a wide range of technologies and concepts included in this list. But the most important, and in fact, the key to making very significant reductions in labor time, is knowledge-based systems. The other technologies, especially database systems and communication networks, can provide good benefits, but KBS has the greatest *potential* of changing the E&C industry and other similar industries, assuming that we can really capture the knowledge for the numerous things involved in our projects, and keep that knowledge up-to-date.

Object Oriented Programming (OOP) method and technologies have a special appeal because so much of the work that goes into plant construction are physical objects that should be amenable to OOP object representation. Considerable effort was expended in developing object oriented concepts and applications, and we looked at how a very large object knowledge base could be developed and maintained. We also developed other types of KBS applications, including a variety of rule-based expert systems. But because of the huge number of objects involved in even a small project, the development of that knowledge base did not continue. The concept is valid and holds much promise, but a way will have to be found to reduce the costs of development, and maintenance of the knowledge base.

In order to understand the very large number of possible ways KBS systems can be applied to E&C projects, consider the list of possible KBS applications shown on Figure 7.5. Note that every area and every discipline has the potential for using KBS applications. Also, some of these involve 3-D CAD models, while others work with schedules, engineering calculations, and database programs. The functions these systems perform include: layout, planning, routing, sizing, selecting, estimating, and most of the other things that people do with data and information.

It is estimated that there could be *several hundred* significant KBS applications and many thousands of objects for a large E&C project. And, while some of the systems, approaches, and methods used on one application can benefit the development of other applications, much of the knowledge is different, and therefore, must be captured for each application. For example, the design of a piping system for an oil refinery is totally different than for a semiconductor fabrication plant.

We also found that the costs of developing the KBS applications and the OOP object knowledge base were much greater than earlier estimated. Although we never got to the point of setting up a software "production" capability, we did have some very good people working on this project, and we were using several other companies to develop certain specialized KBS applications. The costs of the applications we

- Semi Automatic Plant Layout
- Piping Specification Advisor
- Process Instrumentation Advisor
- Automated Pipe Router
- Automated Conduit & Cable Router
- Automated Structural Design System
- Chemical Hazards Advisor
- Semi Automatic Project Planning & Scheduling
- Semi Automatic Piping & Instrumentation Diagram
- Transportation & Shipping Advisor
- Automated Vessel Design System
- Cost Estimating Advisor

FIGURE 7.5.   List of possible KBS applications.

developed ranged from approximately $10,000 to $1,000,000, with most being in the order of $100,000.

In addition to the development costs, there are also maintenance costs. We found that these knowledge bases had to be continually updated and changed more than traditional software so that maintenance costs are likely to be higher.

The conclusion was that the grand vision we had hoped to accomplish was just too expensive and risky. There were too many unknowns and potential pitfalls that could have negated a very large investment. So, in the end, we have decided not to pursue the ultimate approach, but rather to pursue systems that we know will work and will make financial sense.

But, if this "grand vision" could be done (or when it is done), the benefits can be enormous. First, the number of home office personnel on the project can be reduced significantly, perhaps by as much as 80 to 90%. The few that remain will be either the experts who are needed for the creative, innovative, and new elements of the facility, or the generalists who know a number of different disciplines, and who know how the systems work. They will participate in design decisions, and they will monitor the KBS operations and results.

In addition to the very high reduction in labor costs, the engineering activities can proceed much faster. Under the method used today, it takes months to complete the design detail on a project, and it is expensive to look for multiple alternatives. With these new design systems, it may be possible to complete the design in a matter of days, and to be able to consider multiple design alternatives.

As indicated above, we are no longer looking for the total system, but we are still working on methods and KBS systems. Within Fluor Daniel, a few KBS applications are working with several more being developed. Other engineering companies are likely working on similar applications and concepts, and several software companies are now selling KBS products for certain engineering functions.

IMPLEMENTATION ISSUES

There are numerous technical books about designing KBS systems, including ones on the underlying technology and on knowledge representation and capture, but most of the time, the really difficult problems — the ones that determine success or failure of a development project — are not technical. Rather, these problems deal with the cost and schedule of development, the usability and acceptance of the application, and the benefits and payback from using the applications.

Based on the experiences of the author, the following are some of those important implementation issues th at need to be carefully addressed. First, in developing the KBS system, the overall work process of using the application must be considered. If the application is to be beneficial, the users must see a substantial reduction in work hours and schedule time. However, if it takes longer to get ready to run a KBS application, or to utilize the results, then people will not use it. In most cases, the KBS will need a variety of information and data, some that is already in a database and some that has not been captured in a database. Therefore, a user will have to manually enter that data. If the manual entry is too long and difficult, relative to the function performed, the system will not be accepted.

Therefore, the implementation of various applications should start with the applications that are used at the beginning of the project and proceed in sequence as the work process of the project is performed. This method works much better because, in the beginning of a project, the amount of information and data is relatively small so there will not be a lot of manual entry. Then the output will be in a database ready for the next application.

The second factor in "successful implementation" is the organizational acceptance of a KBS, especially expert systems, by a group of people who consider themselves as experts. Therefore, they will likely have differences of judgment and decision-making. As KBS applications are introduced into the work process, ways must be found for getting everyone in the organization to use and support them.

Another factor in the implementation of KBS technology is dealing with the many differences that occur in the knowledge base for applications that are slightly different. For example, the "knowledge" for the structural design of a chemical plant is different than for a food plant. The basic engineering principles do not change, but the knowledge regarding the differences in industry-accepted approaches is different. This implies that there will need to be many more different versions to handle the variations due to different industries, locations, client preference, etc. A major failing of current KBS technology is the costs of dealing with these small variations that can add up to be overwhelming.

Finally, the cost and effort required to keep the knowledge up-to-date was more than expected. Domain knowledge changes, new products are introduced, and new domain experts are monitoring the systems, all of which creates the need to keep working on the knowledge base. In the general theory of KBS, an application is expected to begin as only moderately knowledgeable (hopefully better than most users), but as time goes by and more knowledge is added, the KBS application should become better and better. While this concept may work for some applications, the rapidly changing nature of the E&C business indicates the cost will be high.

CONCLUSION

Achieving financial successes for KBS systems in the E&C industry has been difficult, that is, in the traditional terms of making profits based on the value of revenues less costs. Part of the problem is related to measuring the revenue side. The costs are relatively easy, but revenues related to having and using KBS applications are more difficult to measure accurately. The situation is analogous to the introduction of CAD systems. During the early years of CAD systems, it was not clear that those systems were saving money and paying back the investments, but it just seemed like CAD systems had to eventually save time and money. Now, there are almost no drafting tables, and it would be almost impossible to design a large project in the U.S. without CAD systems. It is the same with KBS systems. In the short term, there will be successes and failures of different applications, but in the long run, knowledge capture and reuse just makes too much sense. However, the technology will have to get better and become less expensive, before it will be used for the many complex applications found in the E&C industry.

Finally, there is a warning about protecting the knowledge base once it is developed. Knowledge is very valuable in the highly competitive E&C business. One of the key decision criteria that our clients use in selecting an engineering and construction company is capability, know-how, and experience. As the know-how to make decisions and perform designs become captured in a knowledge base, the competitive advantage becomes subject to be accessed, moved, bought, or stolen. Therefore, these knowledge bases become the true crown jewels of companies. We must be careful of letting this genie out of the bottle by capturing our valuable knowledge in one computer system.

REFERENCES

1. C. L. Dym and R. E. Levitt, *Knowledge Based Systems in Engineering*, McGraw Hill, New York, NJ, 1991.
2. E. V. Berard, *Essays on Object Oriented Software Engineering*, Prentice Hall, Englewood Cliffs, NJ, 1993.

8 CommonKADS: Knowledge Acquisition and Design Support Methodology for Structuring the KBS Integration Process

Robert de Hoog

CONTENTS

INTRODUCTION

In the early days of knowledge-based system (KBS) development, the so-called "research lab" metaphor dominated the way people were thinking about the organizational role a KBS could perform. This metaphor was characterized by an almost exclusive focusing on the knowledge embodied in the system, which was seen as a stand-alone intelligent program that brought to life the ultimate ambition of AI research. Efforts to describe a "methodology" for building KBSs paid attention only to knowledge acquisition and knowledge representation (see, e.g., Waterman, 1986). However, in the harsh reality of non-research environments, this metaphor proved to be totally inadequate. Suddenly one realized that KBSs had to be fielded in an organization that had no *a priori* positive attitude towards them, that KBSs had to

0-8493-3116-1/97/$0.00+$.50

communicate with users and also with other information systems, that KBSs must be developed in projects that had to live with the usual constraints on time, money, and quality. This in turn led to the realization that for developing viable KBSs more was needed than a perspective on knowledge only.

PERSPECTIVES IN SYSTEM DEVELOPMENT

The evolution in thinking about building KBSs followed, in hindsight, the same line as occurred in conventional software development. Among others, Boehm (1988) describes the original practice as the "code-and-fix" approach which consisted of the following steps:

1. Write some code.
2. Fix the problems in the code.

This implied that thinking about requirements, design, test, and maintenance was postponed until later or not carried out at all. There is a striking resemblance with the way the first KBSs were built by the steps *acquire the knowledge* and *code the knowledge*. This is not the place to elaborate the drawbacks of this approach as they are described comprehensively in the literature (see Boehm, 1988, and many others). It suffices to say that all subsequent alternatives to the "code-and-fix" approach, ranging from strictly linear waterfall models to highly iterative spiral models, advocated a broadening of the scope of the development process by including more perspectives than only the code-perspective. There seems to be an agreement that a comprehensive approach to developing software should at least cover the following perspectives:

1. The *organization* perspective reflects the structural properties of the organization, including aspects like work satisfaction, employment, and distribution of work.
2. The *communication* perspective focuses on the communication channels and communicative behavior surrounding the system to be developed.
3. The *goal/function* perspective zooms in on the key functions an organization has to perform in order to survive and the information infrastructure needed for its survival.
4. The *process* perspective shows the sequence of business processes needed to produce the products and/or services delivered by the organization to its environment.
5. The *data* perspective investigates the data and the relations between the data which are crucial for the goals and functions of the organization.
6. The *change* perspective looks into the changes which are necessary to field a system successfully in an organization.
7. The *project management* perspective covers the planning and control of the process of building the system.
8. The *software engineering* perspective is concerned with the designing and writing of the code in such a way that is satisfies certain quality criteria.

It is not the purpose of this chapter to review the tremendous amount of "methods,"* "methodologies," "approaches," and "techniques" spawned by theorists and practitioners since the emergence of the first modifications on the "code-and-fix" approach. Surveys and comparisons carried out, for example in The Netherlands (see NGI, 1986; NGGO, 1990) include at least 16 "different" methodologies, omitting 20 more. The most important findings of this and comparable reviews can be summarized as follows:

- Most methods omit the organization perspective.
- Techniques are very much the same across methods that profess to be different.**
- Terminology differs widely, but on closer inspection very often the same concepts are meant.
- Most methods are not built on a theory, but a theory could be found after the method had been used with a certain amount of success.
- Notwithstanding the blooming of methods during the last two decades, the performance of the software industry in terms of delivering against time, budget, and quality is still fairly dismal.***

The development of methods and methodologies tailored to the peculiarities of building KBSs followed, as has been said already, more or less the same pattern. This chapter will focus on a concise description of one of the existing methodologies for building KBSs: CommonKADS. It will be shown that by combining the perspectives mentioned above, CommonKADS can justifiably lay a claim on being a comprehensive and consistent methodology for building and fielding KBSs that will greatly facilitate the KBS integration process.

PERSPECTIVES AND MODELS IN COMMONKADS

There are quite a few good descriptions of the CommonKADS methodology available,**** and it is not the goal of this paper to provide a detailed account of it. The objective is to elaborate the often implicit perspectives that underlie it. Hopefully this will show that, contrary to what seems to hold for most methodologies (see above), the "theory" preceded the methodology.

---

* I will not try to define the subtle differences between the subsequently mentioned terms, as most authors fail to do this also. The reader can be referred to Wielinga et al. (1994) for a (not commonly accepted) definition of them.

** A very nice example is the way processes are depicted in Data Flow Diagrams. Yourdon (1989) draws a process as an "oval" while in SSADM (which is widely used in the U.K.) a process is drawn as a "rectangle" (Skidmore, 1994). There is no reason for this difference other than the need to be "proprietary" or even "new" and "unique."

*** There is even some evidence that working according to a method or methodology *decreases* the chances of running a successful project (see Butler Cox, 1990).

**** The best reference for the original KADS is (Schreiber et al., 1993). The enhanced version CommonKADS is described (Schreiber et al., 1994a), though a self contained book is still missing. Most of the project reports can be obtained through the Internet from the web site http://swi.psy.uva.nl/projects/Common KADS/Reports.html.

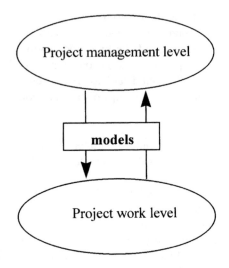

FIGURE 8.1.   Project management level and project work level perspectives.

The first and probably most important distinction made is between *work* and *control of the work*. Though this is familiar to most people in organizational theory, the distinction still seems to elude most practitioners in software development (see Berkeley et al., 1990). In CommonKADS, a rigorous conceptual distinction is made between them, leading to two main perspectives:

- the *project control* perspective, embodied in the project management level
- the *project work* perspective, embodied in the project work level

It is crucial that both levels are connected. The planning and control from the project management level should have "hooks" on the ongoing work at the project work level. The concept that enables this coupling is a *model*, to which I will return later. In Figure 8.1 it is shown how both levels are connected.

Figure 8.1 is of course not very informative. The general concepts must be fleshed out, and that is where other perspectives appear. Let me start with the project management level.

PERSPECTIVES AT THE PROJECT MANAGEMENT LEVEL

The task of the project management level is to control the project work in such a way that the products to be delivered are delivered on time, within the budget while satisfying well-defined quality criteria. These conditions are common to all projects and the specific flavor of CommonKADS is the way the process of project management is modeled. For this, we owe much to the risk-driven approach advocated by Boehm (1988). Use is made of three perspectives that drive the project management process:

FIGURE 8.2.   The CommonKADS project management cycle.

1. *objectives* that must be reached in a cycle,*
2. *risks* that must be addressed in a cycle, and
3. *quality* that must be achieved by products produced in a cycle.

As is described in more detail in de Hoog (1995), the work to be carried out at the project work level is defined by the objectives one wants or needs to pursue, the risks that are threatening the achievement of the objectives, and the required quality of the products. As there are mostly more cycles in a project, these perspectives recur in every cycle. The model of the project management process, the project management cycle, that results from these perspectives is shown in Figure 8.2. Figure 8.2 is a specification of the project management level in Figure 8.1.

Below, a cursory description of the main steps in the cycle in Figure 8.2 is provided.

- **Review** — The most important thing to do is to identify cycle objectives. These objectives can be based on goals negotiated with the client, a basic project plan that acts as a first approximation of a cycle plan or more cycle plans, or on specific threats already identified. This leads to an initial identification of products to be delivered in a development cycle.

---

* A cycle is a limited period in the project life for which a plan is made and whose execution is controlled.

- **Risk** — In this segment the threats to the achievement of the objectives (quality threats) or the realization of the products are analyzed. As a result of the risk analysis, the products receive a priority or new products are added.
- **Plan** — The products identified in the previous segments must be realized by project work. It is assumed that this work can be done in chunks that traditionally are called activities or tasks. The first step in this segment is to define the activities that will realize the products. As soon as these activities are known estimation of effort and time, resource allocation and planning the time scale can be carried out.
- **Monitor** — Work is carried out in the project and this work is monitored to deal with unexpected events that can jeopardize the plan. The products produced by the activities are checked against the quality criteria defined for the products. If the product satisfies the criteria, it is accepted; if not, it will need reworking. Then it can be an input to the next cycle as an objective ("redo the work") in the "Review" segment.

Summarizing the perspectives driving CommonKADS at the project management level, it can be stated that they cover completely the *project management* perspective mentioned in the previous section. Additionally, through the "sub"perspectives objectives, risks and quality, the *change* and *goal/function* perspectives from the previous section are included to a certain extent.

Before turning to the project work level, more must be said about the notion and the role of models in CommonKADS.

Models in CommonKADS

In Figure 8.1 a model is shown as the concept that connects the project management and project work level. Generally speaking, a model in CommonKADS serves three purposes:

1. For the *client,* it is the product that will be delivered to satisfy the requirements.
2. For the *project manager,* it is the product whose production must be planned and controlled.
3. For the *"project worker,"* it is the product that must be worked on.

As there are models at several levels of abstraction and specificity in CommonKADS, the best thing to do is to start with the most general description of a model. In Figure 8.3, this is presented in a kind of entity-relation modeling representation. The boxes are the entities, the annotated links between the boxes are the relations.

The dashed line in Figure 8.3 is of particular significance because it separates the "upper" part of the model structure from the "lower." This is not an accident. The "upper" part is the hook with the project management level because it contains the entities and relations that are relevant from that angle: states, milestones, depen-

Management level

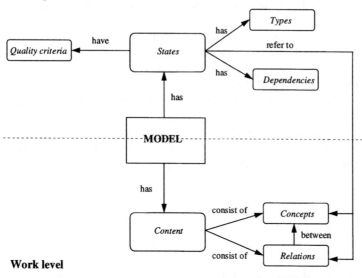

Work level

FIGURE 8.3.   The general model structure in CommonKADS.

dencies, and quality. In the previous section, it was shown that these concepts are needed in the project management cycle (see Figure 8.2). The "lower" part contains the hook with the project work level because it "defines" the concepts and relations that must be developed in order to obtain a (model) product. However, a critical observer will remark that there is not much "defined" yet because the entities concept and relation in Figure 8.3 are as empty as the project management level in Figure 8.1, until a more precise definition was given in Figure 8.2. Thus, there is a need for a more detailed specification of what is meant with these general terms. In doing this, the nature of the project work level is specified simultaneously because the goal of this level is to build the structures described.

Faced with the task of specifying concepts and relations from Figure 8.3, at least three approaches can be followed which reflect increasing commitments to a specific domain.

1. Provide a *notation* or *language*. This is the road that has been taken by most methodologies for conventional systems. A plethora of notational conventions and semi-formal languages are available for expressing concepts and relations between concepts. They all represent a weak ontological commitment to a domain. Data Flow Diagrams as a notation, for example, only presuppose the existence of such general categories as "processes" and "incoming" and "outgoing" data flows, which respectively are slightly more specific descriptions of "concepts" and "relations between concepts." The weak commitment to a domain shows because without any significant modifications to the language one can also model other systems that consist of processes and flows of "matter" (e.g., a

refinery). This approach benefits from its generality, but the hard part for the modeler is to find the empirical counterparts for these general concepts.*

2. Provide a set of *model templates.* A stronger commitment to a domain is realized by defining generic model templates which more clearly reflect observable empirical counterparts. For example, one could make the notion of a "process" more precise by defining the type of processes one can encounter during modeling. Another example is the specification of the notion of a task by prescribing which features of a task can be found (e.g., frequency, knowledge intensiveness, load, etc.). A more specific relation could be that the model template stipulates that in an organization people *derive* power from knowledge. Compared to the first approach there is a loss of generality: concepts like people and relations like derive are more specific than "process" because people cannot arbitrarily be replaced by "animals" or another category, without losing the meaning of the model template (see later).

3. Provide a *reference model.* A ready made model for a particular domain is available. This needs only minor editing and modifying to let it fit the empirical situation under consideration. Examples are reference models for hospitals, for maintenance organizations, etc. They benefit from the large amount of domain information they contain, but they pay the price of decreased generality. If you are not trying to model a hospital but a command center of an aircraft carrier, a hospital reference model will be of little use while Data Flow Diagrams may still help in describing the communication flows.

In a methodology, a mixture of all three approaches can occur and CommonKADS is no exception to this. However, the models that fill the "models" box in Figure 8.1 are all defined as *model templates.* They contain generic concepts that refer to important elements in developing KBSs, but at the level of the instantiation of these concepts in a project quite often a modeling language of notation is provided by the methodology. The CommonKADS models (or model templates) embody different perspectives on a KBS which to a large extent are the same as mentioned in the "Perspectives in System Development" section. Thus, the models can be seen as perspectives that drive the project work level.

Perspectives at the Project Work Level

The full set of model templates in CommonKADS is described in de Hoog et al. (1994). It is outside the scope of this paper to go into them in detail. Below I will succinctly characterize them and will show how they relate to the perspectives.

---

* Teaching this approach shows that the notational conventions can be learned very quickly but as there are no things in the real world that carry a sign "I'm a process" or "I'm an entity," it takes quite some time for novices to learn to identify the "right" processes and entities. The same holds for almost all other notations in this category, including object orientation.

- **The organization model** — The organization model (see de Hoog et al., 1996) addresses issues concerning the socio-organizational environment of the KBS. It contains a number of different views on the organization or part of the organization that enable the people involved to identify problems areas where KBSs could be fruitfully introduced, but also areas where problems can arise when an actual system is put into operation. Examples of these views are the process view, the authority/power view, the functional view, the structural view, and the knowledge view. This model encompasses several perspectives mentioned in the "Perspectives …" section, the most important being the organization perspective, the goal/function perspective, and the change perspective.
- **The task model** — The task model (see Duursma et al., 1993) is a specification of how an organizational function can be achieved by means of a number of tasks, with a focus on knowledge-intensive tasks. This is achieved by means of a task decomposition. Furthermore, there is a description of a number of important properties of tasks like inputs and outputs, capabilities needed to carry out the task, task frequency, etc. For one or more knowledge-intensive tasks, the required expertise is modeled in the expertise model. From the perspectives mentioned in a previous section, the process perspective is covered by this model
- **The agent model** — This model (see Waern and Gala, 1993) describes the important properties and capabilities of agents* that perform tasks in the context of the KBS. This information is important for deciding about the task assigment in the new situation. It also models the communication capabilities of agents, which provides important inputs for the communication model. There is no perspective uniquely associated with this model, though it deals with aspects of the organization.
- **The expertise model** — This is one of the central models in the CommonKADS methodology as it is one of the main components that distinguishes KBS development from conventional system development. It contains a specification of the problem-solving expertise required to carry out the expert's task. The structure of this model is three layered: the domain layer, the inference layer, and the task layer (not to be confused with the task model). It is outside the scope of this paper to describe this important model in more detail; the reader can find them in Schreiber et al., 1994a and Schreiber et al., 1994b.** Clearly, the expertise model introduces a perspective that is absent in the section about perspectives: the *knowledge* perspective.
- **The communication model** — The communication model (see Waern et al., 1993) is devoted to an issue that is frequently overlooked in KBS development: the interaction with the user and other software systems

---

* The term "agent" refers to any entity that can carry out a task. Thus, an agent can be a human, but also could be a computer or another machine.

** Schreiber et al. (1994b) is an example of how "inside" a model a notation or a language can be provided for the modeller. The interesting thing is that "inside" the expertise model lower level templates are also available for generic reasoning processes (see Breuker & van de Velde, 1994).

(e.g., databases). Though this model is concerned with the communication perspective, its scope is less broad than indicated in the perspectives section. It only covers human-computer and computer-machine communication and not person-person communication even if it is relevant for the context of the KBS.

- **The design model** — The bridge to actual implementation of the system is provided by the design model (see van de Velde et al., 1994). This consists of a description of the computational and representational techniques that must be used for building and running the KBS. Its role is straightforwardly related to the software engineering perspectives.
- **The system model*** — This is the software, the KBS that has to carry out all or some of the knowledge-intensive tasks modeled in the expertise model and has to operate in an organizational context defined by the organization model. There is no particular perspective associated with this model though in constructing it software engineering principles will play a role.

After characterizing the models in CommonKADS and how they are related to the perspectives in the earlier section, it can be shown how they realize the integration of KBSs. But before pursuing this goal I will return to a topic raised in the "Models in CommonKADS" section. It was argued that CommonKADS provides a set of model templates as a way of dealing with the need to specify the concepts and relations in Figure 8.3. In Figure 8.4, the model template of the agent model is given to show how this is achieved.

The agent model is chosen as an example because it is one of the more simple models in CommonKADS which can serve the explanatory role best.

In Figure 8.4, the model elements are the rounded boxes and the annotated arrows are the relations between model elements (see also Figure 8.3). It is easy to see that "element" has been specified into four separate ones: the agent and its properties (name, type, etc.), the constraints on the agent (norms, preferences), the general capabilities of the agent (general skills, etc.), and the reasoning capabilities. These four elements tell the developer more about what has to be modeled from an agent, if the actual instances of these elements will be different for different projects or within the same project (the agent model can contain more instances of the model template). The model relations are specified likewise. From the model template, we can see that reasoning capabilities of the agent are described in the expertise model, that the agent participates in transactions between the agent and the KBS, and that the agent performs a task and receives and supplies ingredients (inputs, outputs).

Figure 8.4 can also be used to illustrate the "refer to" link in Figure 8.3. Recall that this link couples the management level and the work level. Thus, states that need to be planned and monitored at the management level are described in terms of the model elements and relations they refer to. In terms of Figures 8.3 and 8.4,

---

* Though this model is not part of the original model set defined in de Hoog et al. (1994), experience in using CommonKADS has shown that there is definitely a need to be able to represent the software as a separate model.

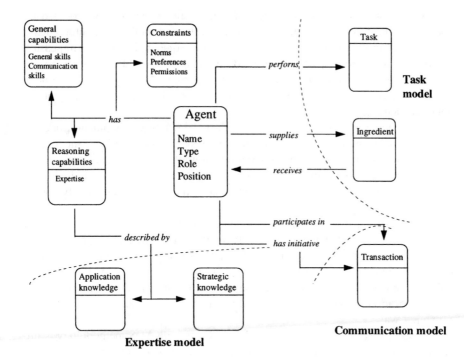

FIGURE 8.4.   The CommonKADS agent model.

the plan for a cycle could contain a statement like: "describe the norms, preferences and permissions of agent X." This task has to be carried out at the work level, that is finding "values" for norms, preferences, and permissions. The results ("values") are reported back to the management level and a quality check is carried out (the "assess quality" step in Figure 8.2).

There is another interesting aspect in Figure 8.4. It contains three dashed lines which represent boundaries between models. However, some relations "cross" these boundaries. This brings me back to the main theme of the paper: integration.

INTEGRATION

Though the CommonKADS models were described separately in the "Perspectives at the Project Work Level" section, their crucial feature is that they are all interrelated by means of well-specified links as indicated in Figure 8.4. This linking of models forces the developer to consider the broadest context that will facilitate the KBS integration process. The success of a KBS will depend to a large extent on how in the design and development process the important perspectives are taken into account. In CommonKADS, this means that if you work on a particular model you must *always* consider the links the model has with other ones. For example, in Figure 8.4, modeling an agent implies that you *have* to think about the task(s) the agent has to perform and the communicative transactions the agent will become involved in, which automatically brings in the task and agent model. This works also the other way: when working on the task or communication model, you automatically encounter the links with the agent model. As a consequence it is almost impossible to *overlook* the relevant aspects in KBS development and fielding. This is not the same as not paying attention to it, because it is permitted to omit model elements if there is neither an objective nor a risk that can be addressed by working on the model element(s).

The second aspect of integration is reflected in how development work and project management are coupled. They *both* address the same products, the models, and their integration consists of a continuous flow of model building assignments from the project management level to the project work level and a flow back consisting of model building results. Additionally, the often elusive quality control activity fits naturally in this process.

Summarizing, the CommonKADS methodology provides a comprehensive and consistent way to contribute significantly to the KBS integration process. It is comprehensive as it covers the most important perspectives which are relevant for KBS development. It is consistent because it encapsulates a close coupling between project management and project work expressed in basicallly the same model concepts. Integration is more or less enforced by the way the models are interrelated in an explicit way.

The substantial integrative potential of the CommonKADS methodology is further illustrated by the fact that parts of it can be and are (re)used for guiding knowledge management work,* which requires a very broad perspective on the role knowledge plays in an organization (see van der Spek and de Hoog, 1995).

REFERENCES**

1. Berkeley, D., R. de Hoog, and P. Humphreys, *Software Development Project Management: Process and Support.* Ellis Horwood, 1990.
2. Breuker, J. and W. van de Velde, *The CommonKADS Library for Expertise Modelling.* IOS Press, Amsterdam, 1994.
3. Boehm, B.W., A spiral model of software development and enhancement, *IEEE Computer,* May issue, (61–72), 1988.
4. Butler Cox, *Trends in Systems Development,* 1990.
5. Duursma, C., O. Olsson, and U. Sundin, *Task Model Definition and Task Analysis Process.* Deliverable ESPRIT Project P5248, KADS-II/M5/VUB/RR/004/02, Free University of Brussels, Brussels, 1993.
6. Hoog, R. de, R. Martil, B. Wielinga, R. Taylor, C. Bright, and W. van de Velde, *The CommonKADS model set.* Deliverable ESPRIT Project P5248, DM1.1c, KADS-II/M1/DM1.1b/UvA/018/6.0/final. University of Amsterdam, 1994.

---

* Projects in this area include knowledge mapping and knowledge modeling for the Royal Dutch Navy, improving the internal and external knowledge market for a Faculty of the University of Amsterdam, and increasing the knowledge exchange flows in a large internationally active food supplier.
** See http://swi.psy.uva.nl/projects/CommonKADS/Reports.html for project reports.

7. Hoog, R. de, "Project management in CommonKADS: integration of products, objectives and risks". In: F.J. Cantú-Ortiz, R. Soto, M. Campbell and J.M. Sánchez (Eds), *Proceedings 8th International Symposium on Artificial Intelligence*, ITESM, México, (27–35), 1995.

8. Hoog, R. de, B. Benus, M. Vogler, and C. Metselaar, "The CommonKADS Organization Model: content, usage and computer support." *Expert Systems with Applications,* Vol. II(1), 29–40, 1996.

9. NGGO, *16 methoden voor systeemontwikkeling.* (16 methodologies for system development). Tutein Nolthenius, Amsterdam (in Dutch), 1990.

10. NGI, *Methodieken voor informatiesysteemontwikkeling.* (Methods for information systen development), Technical Report NGI, Amsterdam (in Dutch), 1986.

11. Schreiber, G., B. Wielinga, and J. Breuker (Eds.), *KADS. A Principled Approach to Knowledge-Based System Development.* Academic Press, New York, 1993.

12. Schreiber, G., B. Wielinga, R. de Hoog, H. Akkermans, and W. van de Velde, CommonKADS: a comprehensive methodology for KBS development, *IEEE Expert,* Vol. 9 (6), (28–37), 1994.

13. Schreiber, A.Th., B.J. Wielinga, J.M. Akkermans, W. Van de Velde, and A. Anjewierden, "CML: the CommonKADS conceptual modelling language". In: L. Steels, A.Th. Schreiber, and W. Van de Velde (Eds.), *Proceedings European Knowledge Acquisition Workshop EKAW'94*, vol. 867 of *Lecture Notes in Artificial Intelligence,* p. 1–15, Springer Verlag, 1994.

14. Skidmore, S. *Introducing Systems Analysis.* NCC Blackwell, 1994.

15. Spek, R. van der and R. de Hoog, "A framework for a Knowledge Management Methodology". In: K. Wiig, *Knowledge Management Methods.* Schema Press, p. 379–393, 1995.

16. Van de Velde, W., C. Duursma, G. Schreiber, P. Terpstra, R. Schrooten, V. Golfinopolous, O. Olsson, U. Sundin, and M. Gustafsson, *Design Model and Process.* Deliverable ESPRIT Project P5248 KADS-II/M7/VUB/RR/064/2.1, Free University Brussels, University of Amsterdam, Swedish Institute of Computer Science, Cap Programator, Brussels, 1994.

17. Waern, Y. and S. Gala, *The CommonKADS Agent Model.* Deliverable ESPRIT Project P5248 KADS-II/M4/TR/SICS/002/V.2.0, Swedish Institute of Computer Science and ERITEL, Kista, 1993.

18. Waern, Y., K. Höök, R. Gustavsson, and R. Holm, *The CommonKADS Communication Model.* Deliverable ESPRIT Project P5248. KADS-II/M3/TR/SICS/006/2.0, Swedish Institute of Computer Science and ERITEL, Kista, 1993.

19. Waterman, D. A., *A Guide to Expert Systems.* Addison-Wesley, 1986.

20. Wielinga, B.J., A.Th. Schreiber, and R. de Hoog, "Modelling Perspectives in Medical KBS Construction". In: P. Barahona and J.P. Christensen (Eds.), *Knowledge and Decisions in Health Telematics,* IOS Press, p. 95–102, 1994.

21. Yourdon, E. *Modern Structured Analysis.* Prentice-Hall, 1989.

9 Corporate Downsizing: Preserving the Core Knowledge Base by Utilizing Expert Systems, Strategic Decision Support Systems, and Intelligent Agents

Mohsen Modarres and Ali Bahrami

CONTENTS

INTRODUCTION

To establish greater internal efficiency and effectiveness, more than 80% of the Fortune 1000 corporations have strategically downsized their white-collar staff (Cameron, Freeman, and Mishra, 1991). A major portion of such reductions has been in response to technological change (Anderson & Tushman, 1990) and declining white-collar productivity. Moreover, for the past decade fundamental changes

0-8493-3116-1/97/$0.00+$.50
© 1997 by CRC Press LLC

in the competitive forces in the marketplace, increased foreign competition, corporate raiding, and increasingly shrinking resources have pressured major corporations to restructure through reengineering (Daft, 1995), elimination of managerial layers, and downsizing the portfolio of their businesses (DeWitt, 1993; McKinley, 1993). Recently, there have been significant work reductions within large and complex corporations (e.g., G.M., IBM, and Boeing). However, a report by the Wall Street Journal has suggested that few corporations may have achieved the downsizing goals of reduction in expenses, higher productivity, and gaining a competitive advantage (Hitt, Ireland, and Hoskisson, 1995).

The recent resurgence of corporate re-engineering and downsizing strategies has been spurred by an interest in assessing the dynamic and selective aspects of downsizing strategies and the long-term impact of such strategies on managerial functions and capabilities and overall corporate performance. Moreover, research on the structural and performance consequences of strategic downsizing have not provided additional criteria (DeWitt, 1993) for the retention of the knowledge base within corporations and to enhance top management's capabilities to be downsized selectively (Hardy, 1986). The complex and multifaceted nature of corporate restructuring, however, has created a healthy debate among both researchers and practitioners as to whether downsizing strategies enhance the long-term survival and competitive position of corporations or caused the disruption of routine activities, communication channels, and loss of knowledge base and corporate competence. As such, the purpose of this chapter is to examine how information technologies can facilitate downsizing without losing the core corporate knowledge base and expertise.

CORPORATE DOWNSIZING

Downsizing constitutes a set of activities implemented by top echelons of organizations in order to improve structural design, operative efficiency, and corporate competitiveness (Freeman and Cameron, 1993; Hitt, Ireland, and Hoskisson, 1995). As such, it represents strategies that affect the size of the corporate work force and the work processes employed (Freeman and Cameron, 1993). Corporate downsizing may also be focused on greater efficiency or effectiveness. Bruton, Keels, and Shook (1996) remarked that when a firm downsizes, it reduces the number of employees. According to the authors, the consequences of employee reduction may be greater efficiency, as the firm carries out the same functions with fewer employees, or effectiveness as firms divest subunits and departments and abandon certain industries.

Downsizing is both a dynamic and selective process. The dynamic aspect of downsizing is concerned about the magnitude of change. For instance, corporations may engage in drastic revolutionary cuts and simultaneously redesign the corporate structure. Within the selective processes the major concern tends to be how to downsize and, at the same time, retain the knowledge base of the corporations. The dynamic and selective processes of downsizing tend to pressure the top executives to use the processes strategically. That is, success in implementing strategic downsizing tends to be contingent on the magnitude of downsizing , and the top managers' capabilities in maintaining core skills, and performing their functions with fewer employees and managerial layers. Moreover, success of downsizing tends to be contingent on the establishment of vertical communication mechanisms that can facilitate the efficient flow of information, and strategically link top decision-makers with lower levels of the corporation.

MAGNITUDE OF CORPORATE DOWNSIZING

Corporate downsizing may vary in magnitude. Industry conditions and the financial health of corporations influence the magnitude and scale of cuts. Corporate redesign and downsizing may be accomplished through elimination of peripheral activities and subunits and reallocation of resources toward enhancement of core competencies within the existing managerial and structural framework. Simply stated, corporations can perform their previous functions with fewer employees. Hence, the loss of skills and knowledge from peripheral activities are of less strategic importance to corporations. However, elimination of such activities saves cost without reorganization of top managerial layers and tasks. Moreover, this approach to downsizing focuses the corporate redesign on core competencies and enhances top management's coordinative capabilities and strategic control over internal activities.

Downsizing may be drastic in its magnitude. Greater international competition and resource scarcity tend to stimulate the initiation and implementation of drastic downsizing strategies which involve elimination or merging of departments and reduction in white collar work force (DeWitt, 1993). Drastic downsizing strategies require major reallocation of firm resources as corporations eliminate some activities and reinvest in core capabilities. It may also result in the loss of competencies and knowledge-base as corporations redesign through elimination of mid-level managers with years of work-related experience. For the past decade, large complex organizations (e.g., IBM, Boeing, GTE, and G.M.) have engaged in drastic downsizing of their work force to remain competitive. IBM, for example, narrowed the scope of its operations by remixing its business portfolio and downsized substantially. As a result, close to one million managers have lost their jobs (Hirsch, 1988). According to Hitt, Ireland, and Hoskisson, (1995, 331), "this [downsizing] represents a significant loss of potential human capital [knowledge and skills]. Because of this loss, many firms are operating with little slack in their human capital. It is also not uncommon for restructuring firms to reduce their expenditure/ investment in training and development programs."

Past research has examined the impact of information technology on organizational design (Lucas and Baroudi, 1994), white collar productivity (Davis, 1991), the cost of coordination; the specificity of investment in inter-organization interactions (Malone, Yates, and Benjamin, 1987), and the contribution of middle managers in implementing information technology. This chapter contributes and extends the past research by arguing that expert systems (ES) and intelligent agents are capable of undertaking a major portion of middle managers' duties and will limit the loss of the knowledge base as the size of mid-level management is reduced during corporate downsizing. Additionally, the study argues that expert systems enhance corporate capabilities in responding to market changes in a timely fashion.

TECHNOLOGICAL LEVELING OF MANAGERIAL LAYERS

Information technology is capable of redesigning organizational structure through reducing the barriers between various departments (Robbins, 1994) and various levels of managerial hierarchy. Information technology, which had once been a tool for greater expansion and organizational growth, has become a tool for downsizing (Applegate, Cash, and Quinn Mills, 1988). Information technologies (e.g., ES and strategic DSS) are capable of assuming many of the communication, coordination and control functions that middle managers previously performed (Applegate et al., 1988). These expert systems enhance managers' span of control and reduce functional redundancies in order to group similar decisions under single managers (Robbins, 1994). According to Lucas and Baroudi (1994), information technology can be a substitute for managerial layers and also for a number of management tasks. The author noted that within more complex and bureaucratic corporations, layers of managers (e.g., middle management) primarily disseminate the information from the top management to the lower levels and also provide usable information from the daily operations in order to assist with the strategic decisions by top management.

Restructuring and downsizing complex organizations based on information technology, specifically expert systems, tends to be a challenging task to accomplish. However, the match between the mid-level manager's functions and the expert system's capabilities tends to benefit corporations in limiting the loss of the knowledge base, enhancing strategic decision making (Turban and Watkins, 1986) through electronic links of lower and top management, less bureaucratic layers (Daft, 1995), training and feedback appraisals (Peterson, Van Fleet, and Smith, 1994), and expeditious policy implementation. IDS Financial Services, a subsidiary of American Express, has encoded the expertise of its best accounting managers through expert systems and reported a far better performance by the average planners compared to the planners that did not use such systems (Robbins, 1994).

The first line management is typically responsible for training, supervision of the daily corporate operations, and creation of a comprehensive data base for the top management to draw upon in making strategic decisions. Within the traditional bureaucratic corporations, middle managers have provided usable information to strategic managers for decision making purposes. Hence, first line managers have developed a strong link between the production and operations of various departments and the rest of the organization; and a weak direct link to upper management (Hellriegel and Slocum, 1992).

Mid-level managers, generally, are not strategic decision makers (Stewart, 1987). They are, however, responsible for instructing first line managers in implementation of the corporate policies formulated by top management. The middle managers' skills and knowledge in translating the overall strategies into objectives and plans tends to be critical in successful implementation. The implementation of strategies requires organizing activities, assigning responsibilities, planning the distribution of resources toward assigned activities, and creation of a feedback loop for performance assessments. The knowledge base used in planning, performance feedback, and organizing activities can be accomplished by codifying the knowledge of managers in an ES. The expert system can be used as a training tool for managers [first line managers] in both performance appraisals (Peterson et al., 1994), planning and forecasting. Studies by Putman (1990), and Kearsley (1987) have reported that these systems are already providing managers with information and advice in conducting performance assessments and other managerial tasks. Peterson et al. (1994) also reported that managers who use expert systems tend to perform their task more effectively.

The application of information technology at top management levels may differ from that of first line managers. To exploit information systems in retaining the knowledge base and creating a flatter organizational design, corporate managers ought to understand its application at various managerial levels. Strategic managers are required to have access to usable information on internal capabilities and other factors that impact strategic decision making. Belardo, Duchessi, and Coleman (1994) reasoned that strategic DSS modeling techniques can be instrumental in creating models for product life cycles, product portfolio matrix, and Porter's five forces. According to the authors, these models show the interrelationship among critical models and suggest appropriate strategies. An ES can provide advice and usable information on operational capabilities, and create an interface between the user and the model base. The integration of strategic decision support systems with expert system creates a dynamic and interactive information system that is capable of providing top management with information for specific strategic decisions and first line managers with information for operative decisions. This interactive system reduces the need for the lengthy process by which traditional middle managers would receive the formulated strategies from top managers and translate it into programs, objectives, and plans for first line managers to implement.

PRESERVING THE CORE KNOWLEDGE BY INTEGRATING EXPERT SYSTEMS (ES) AND STRATEGIC DECISION SUPPORT SYSTEMS (SDSS)

Over the last two decades, various approaches have been proposed for improving the decision-making process along with the advancements in computer hardware and software. Expert systems (ES), decision support systems (DSS), and Intelligent agents are the results of these efforts. ES and DSS disciplines contributed to these developments. Although they appear to be different, they share the common goal of developing practical methods for solving problems with different emphases. Both disciplines have much in common, and each has something to offer to the other.

An expert systems approach to problem solving is mostly heuristic, as opposed to the model-based approach of DSS. An expert system is basically concerned with building intuitive systems that resemble a human expert, while DSS is concerned with assisting users in unstructured decision making. Although the inference engine of an expert system provides a general problem-solving capability that can trace the decision tree and explain its activities, it also utilizes what are known as "hard constraints" (Smolensky, 1999). Hard constraints are provided by a knowledge base and set of rules that define the model(s). The conditions which result in the firing of the rules must be noise free in order for the expert system to recognize a match between the environmental events and its *a priori* model (C. Glover and P. F. Spelt, 1990).

An expert system is a domain-specific knowledge base information system that may be employed as an expert consultant to end users (O'Brien, 1994). As such, it can provide top managers with specific knowledge (Geisler, 1986) which can be used in accomplishing their tasks (Kearsley, 1987). Expert systems tend to be domain-specific; the limited problem domain confines the system's functions to defined boundaries (Turban and Watkins, 1986). For example, lower managerial levels typically face limited, well defined problems within a specific domain. The explicit encoding of knowledge can be both a strength and a weakness of an expert system. It is a strength because it can explain its activity and can simplify the development and debugging of applications that are top down or concept driven. On the other hand, "hard constraints" can be weaknesses for a data-driven application with a massive amount of information. Developing and maintaining such a system with an expert system is extremely difficult, since the knowledge engineer must have *a priori* knowledge of every conceivable circumstance that will or might happen (Bahrami and Dagli, 1994). The best one can do when utilizing an expert system is to assess a subset of the problem with highly heuristic rules. At the same time, knowledge-based systems are intrinsically heuristic and are not good at generalization.

Expert systems (ESs) may enhance the task performance of first line management by enabling them to modify the problem structure and speed up routine decision making and problem solving through diagnosis and reasoning capabilities. An expert system also has the capability to link first line management to top management. (See Figure 9.1.) Integrating the expert system and first line management may render several advantages. It can improve accessibility to data and the latest market information to be used in daily operation decisions. It also facilitates the modification of existing problems through dynamic interaction with end users. Moreover, first line managers can benefit from the computerization of decision making processes and the intelligent advice provided by the expert system. Downsizing strategies may exploit the strategic advantage of the expert system and retain the skill and knowledge base of middle managers by employing the expert system which has similar capabilities. Connecting the first line managers with top management through ES enables corporations to merge the duties of mid-level management with first line management, and facilitates the transformation of usable information to strategic decision makers in a timely fashion.

The expert system enables corporations to design a leaner structure with fewer managerial levels by integrating the ES with a strategic decision support system (SDSS). Creating an information system that has the capability to advise, reasoning capabilities, which can provide alternative solutions to implementation problems, can prevent the loss of competence and knowledge base as mid-level managers are eliminated in the restructuring process. As indicated earlier, strategic managers rely on the information gathered by the mid-level managers for strategic decisions on cost leadership strategies, differentiation of products, and competitive positioning of their products and services with respect to other rivals. The integration of ES and

FIGURE 9.1.   An organization can be divided into strategic level and a knowledge operational level. Middle-level management has been replaced by IT to assist line managers with day-to-day activities and strategic managers with unstructured decision making.

strategic DSS facilitates the channeling of appropriate information to strategic mangers as end users. The electronic link between first and strategic management can expedite product modifications and response to market changes without the need for an extra layer of managers to filter the information necessary to make strategic decisions.

Turban and Watkins (1986) argued that synergy can be achieved by integrating expert systems into decision support components. In particular, the synergy can be achieved in strategy formulation and implementations. The authors reasoned that strategic decision making consists of eight steps: specification of objectives, parameters, and probabilities; retrieval and management of data; generation of decision alternatives; inference of consequences of decision alternatives; assimilation of verbal, numerical, and graphical information; evaluation of sets of consequences; implementation of decisions (strategies); and strategy formulation. Turban and Watkins noted that users (e.g., strategic managers) can work with DSS and upon formulation of the strategy the user can call upon an ES. The expert system can conduct further analysis and provide an answer for the strategic managers.

Belardo, Duchessi, and Coleman (1994) argued that strategic decision support systems, as an instance of a DSS, are different from the EIS, which primarily perform information access, monitoring, and control functions for the executives. The authors remarked that in performing strategic decision making tasks, top management can benefit from strategic DSS that present relevant information and models that facilitate a better understanding of external opportunities and threats and internal capabilities and weaknesses. Moreover, the authors noted that these systems can present interdependencies among internal and external factors that are necessary for business strategy development such as internal capabilities, price competitiveness, and product technology. Strategic DSS contain features for strategic analysis.

ES can also complement strategic DSS by presenting explanatory information, facts, and reasoning capabilities. Clearly, the strength of one complements the weakness of the other, since they focus on opposing ends of the problem-solving strategies. By taking advantage of these complementary strengths and weaknesses of expert systems, and DSS, the proposed hybrid system is an attempt to build a more robust system that can perform better than any of its subparts acting alone.

PROVIDING THE LINK: INTELLIGENT AGENTS

Agents are basically human-delegated software entities or personal software assistants that can perform a variety of tasks for their human masters (Cheong 1996). Intelligent agents with sophisticated analysis functionality, such as Data Surfing, and complex filtering for providing strategic managers access to information for analysis, presentation, integration, and action, can aid corporations during downsizing by performing most of the functions of the middle-level managers. Intelligent Agents can provide an analytical power to be distributed through the organization by providing managers with a palette of multidimensional objects, making advanced data mining and ad hoc analysis capabilities directly against relational databases possible. By utilizing these objects, organizations can create DSS applications of unmatched sophistication and flexibility.

An agent can contain features that can enhance the first-line managers' ability to conduct data mining. For example, the Agent's data mining features allow organizations to refine large, unmanageable sets of data into meaningful groups of information, which can then be used for identifying trends, buying habits, and customer behavior.

The Agent reports can be defined to initiate statistical analysis, scenario analysis, production planning, and other applications, allowing the strategic manager to not only access the other applications, but to use the most current data from their corporate database. These intelligent agents can be customized to allow strategic managers to dynamically select filtering criteria for a specific report. This is ideal for the less technical strategic manager.

The Agent's mail enablement provides corporations with complete integration with E-Mail. Reports can be sent directly from the Agent through electronic mail to an individual or to a distribution list of individuals, automating the dissemination of information throughout the organization. Corporations can leverage the efforts of employees by enabling them to share a variety of reports. A report can be sent in both spreadsheet and word processor formats. According to Belardo et al. (1994), Strategic DSS enable top managers to conduct the strategic SWOT analysis (strength, weaknesses, opportunities, and threats) by generating and evaluating strategies, and providing performance feedback. These systems can also explain concepts, stimulate thinking, and combine conceptual models (e.g., value chain models). Moreover, rather than waiting for reports by mid-level managers, strategic managers will be less insulated from the operations through direct access to corporate or external databases which present them with usable information on internal competencies in terms of operational cost, pricing, and manufacturing capabilities, which can be used in business strategy formulation.

BEHAVIORAL ISSUES IN CORPORATE DOWNSIZING

Large and complex organizations contain a greater number of departments, occupational specialties, and managerial levels. Elimination of departments and managerial layers during the process of downsizing may lead to shifts in power, coalition formation, and resistance to change. In particular, the departments and individuals that are directly affected by downsizing strategies tend to resist restructuring and support the status quo. As such, relevant information may be excluded from decision-making processes by managers. Therefore, the success in implementation of strategic downsizing, to a great extent, is contingent upon the elimination of managerial bias in inclusion of all the relevant information in the decision making process. Information technology (e.g., Es, DSS, and intelligent agent) enhances the flow of information across various corporate levels and enables strategic managers and top executives to exercise greater discretion in including all the relevant information.

According to Stewart (1987) middle managers may or may not be committed to the corporation. That is, mid-level managers may have greater inclination toward their own self interests. The choices that middle managers make may not be similar to that of top management. This may be true during the downsizing period as information technology is expected to reduce the need for a portion of services and functions performed by middle managers. The information technology attenuates the biases that may exist in implementation of policies, and the flow of needed information to top executives and strategic managers.

CONCLUDING REMARKS

Increased competition in the marketplace, changes in demand, and the need for greater operational efficiency have pressured a great percentage of corporations to restructure their activities through downsizing. Downsizing has become a popular strategy and is universally accepted by major corporations. However, despite the growing popularity of strategic downsizing, its impact on corporate performance and efficiency is unclear. That is, the success of strategic downsizing is contingent on whether corporations are capable of retaining their knowledge base and core competencies while reducing the vertical differentiation through elimination of mid-level managers.

Our research indicates that expert systems are capable of executing a major portion of the duties previously performed by the middle managers (e.g., decision making, providing usable information to the top managers, facilitating implementation of strategies, and problem resolution). Strategic decision support systems are capable of enhancing managerial capabilities in selecting business strategies, thinking strategically, and responding to market changes in a timely fashion. Furthermore, intelligent agents enable the necessary link between strategic and first line managers. The integration of expert systems, strategic DSS, and intelligent agents can be instrumental in both gaining a competitive advantage in the marketplace and in retaining the knowledge base as corporations downsize and become more efficient in their information processing and operations.

REFERENCES

1. Anderson, B. and Tushman, M. L., Technological discontinuities and dominant designs: A cyclical model of technological change. *Admin. Sci. Q.*, 35: 604–633, 1990.
2. Applegate, L. M., J. I., Cash, and D. Quinn Mills, Information technology and tomorrow's managers. *Harvard Business Rev.*, 126–136, 1988.
3. Bahrami, A. and Dagli, C. H., Hybrid Intelligent Packing System (HIPS): An Intelligent Packing By Combining Artificial Neural Networks, Artificial Intelligence and Optimization, *J. Appl. Intell.*, Vol. 4, No. 4, 1994.
4. Bahrami, A. and Modarress, B., Quality Function Deployment A Symposium, Proceedings of Decision Sciences Institute (DSI), Volume 3, page 1266, Miami Beach, Florida, (November 24–26), 1991.
5. Bahrami, A., Designing Artificial Intelligence-based Software, John Wiley & Sons, NY, 1988.
6. Belardo, S., Duchessi, P., and J. R., Coleman, A strategic support system at Orell Fussli. *J. Manage. Info. Sys.*, 10, 4, 135–157, 1994.
7. Bruton, G. D., Kells, K. J., and Shook, C. L., Downsizing the firm: Answering the strategic questions. *Acad. Manage. Exec.*, 38–45, 1996.
8. Cameron, K. S., Freeman, S. J., and Mishra, A. K., Best practices in white collar downsizing: Managing contradictions, *Acad. Manage. Exec.*, 5, 3, 57–73, 1991.
9. Cheong, F., *Internet Agents: Spider, Wanderes, Brokers, and Bots.* New Riders, Indianapolis, Indiana, 1996.
10. Daft, R. L., *Organization Theory and Design.* St. Paul MN: West Publishing, 1995.
11. DeWitt, R. L., The structural consequences of downsizing. *Org. Sci.*, 4, 1, 30–41, 1993.
12. Freeman, S. J. and Cameron, K. S., Organizational downsizing: A convergence and reorientation framework. *Org. Sci.*, 4, 1, 10–29, 1993.
13. Galbraith, J. R., From recovery to development to through large scale changes. In Morman, A. M., Morman, S. A., Ledford, G. E., Cummings, E. E., and Lawler, E. E. (Eds.), *Large Scale Organizational Change.* San Francisco: Jossey-Bass, 62–87, 1989.
14. Geisler, E., Artificial management and the artificial manager. *Bus. Horizons.* 29, 4, 7–21, 1986.
15. Glover, C. and Spelt, P. F., *Hybrid Intelligent Perceptron System: Intelligent Perceptron Through Combining Artificial Neural Networks and Expert System,* SPI, Auburn University, pp. 321–331, 1990.
16. Goodman, P. S., Sproull, L., S., and Associates, *Technology and Organizations.* San Francisco Jossey Bass Publishers, 1990.
17. Hardy, C., Strategies for Retrenchment: Reconciling individual and organizational needs. *Can. J. Admin. Sci.*, 3, 275–289, 1986.
18. Hellriegel, D. and Slocum, J. W., *Management* (6th ed). Addison Wesley. New York, 1992.
19. Hirsch, P., *Pack your Own Parachute.* Reading, Mass: Addison Wesley, 1988.
20. Hitt, M. A., R. D. Ireland, and R. E. Hoskisson, *Strategic Management: Competitiveness and Globalization.* West Publishing. New York, 1995.
21. Kearsley, G., Software for the management of training. *Training News.* 8, 9, 13–15, 1987.
22. Kochan, T. A. and Useem, M., Transforming organizations. Oxford University Press. New York, 1992.

23. Leidner, D. E. and J. J. Elam, Executive information systems: Their impact on executive decision making. *J. Manage. Info. Sys.,* 10, 4 , 139–155, 1994.
24. Lucas, H. C. and J. Baroudi, The role of information technology in organization design. *J. Manage. Info. Syst.,* 10, 4, 9–23, 1994.
25. McKinley, W., Organizational decline and adaptation: theoretical controversies. *Org. Sci.,* 4, 1, 1–10, 1993.
26. Modarres, M. 1996. Reorganizing complex institutions: The contingent effects of structural complexity and organizational size. Paper in progress. Washington State University.
27. O'Brien, J. A., *Information Systems.* Irwin: Boston, 1994.
28. Peterson, T. O., Van Fleet, D. D., and Smith, P. C., The Application of an expert system to performance feedback task: Empirical findings. Paper presented at Academy of Management, Dallas, 1994.
29. Putman, A. O., Artificial persons. In A.O. Putman and K. E. Davis (Eds.), *Advances in Descriptive Psychology.* 5, 81–103. Boulder Co. Descriptive Psychology Press, 1990.
30. Robbins, S. P., *Management.* Prentice Hall. Englewood Cliffs, New Jersey, 1994.
31. Simon, H. A., Applying information technology to organization design. *Public Admin. Rev.,* 33, 268–278, 1973.
32. Simon, H., *The New Science of Management Decisions.* Harper and Row Publishing. New York, 1960.
33. Smolensky, P., On the Proper Treatment of Connectionism, *Behav. Brain Sci.,* Vol. II, pp. 1–74, 1988.
34. Sutton, R. I. and D'Aunno, Decreasing organizational size: Untangling the effects of money and people. *Acad. Manage. Rev.,* 14, 194–212, 1989.
35. Turban, E. and Watkins, P. R., Integrating expert systems and decision support systems. *MIS Q.* 121–136, 1986.

10 Information and Knowledge Management in Integrated Science and Technology at James Madison University

Richard M. Roberds and Christopher J. Fox

CONTENTS

This chapter is about the implementation of two innovative, forward-thinking ideas: the start of a new college program in science and technology that is revolutionary in concept, and the role of information and knowledge management as an integral part of the program's curriculum. In the first half of the chapter, we deal with the content and approach of the new science program. While the concept of

0-8493-3116-1/97/$0.00+$.50
© 1997 by CRC Press LLC

the program in integrated science is revolutionary and interesting, our chief purpose in presenting it is to outline the vision, intent, and structure of this new paradigm for science education as the basis for explaining how information and knowledge management forms a key element of the curriculum. The second half of the chapter, then, is the heart of the subject.

REFORM OF SCIENCE TEACHING

The idea for a college program in integrated science and technology at James Madison University grew from several quarters. All were concerned with the reform of education in America, and, in particular, the manner in which science, mathematics, and technology were being taught in the nation's school systems.

In the early 1980s, the National Commission on Excellence in Education pointed to the fact that the U.S. had changed from an industrialized society to an information society that demanded reforms in the entire U.S. education system.* In 1985, a year when Comet Halley was making its return to earth after a 76-year absence, the American Association for the Advancement of Science initiated Project 2061. (2061 is the year that Halley will next appear, an event that dramatically demonstrates the compelling interest that science can hold.) Project 2061 was motivated by a concern that without reform in science, mathematics, and technology teaching, young Americans would not be adequately prepared for the challenges of their lives and careers. Grades K through 12 educators responded early to the call for reforms, although it is fair to say that the response has not been as widespread as the reform movement would like to see.

In 1993, the engineering education community took action to implement changes in engineering education. A joint meeting of the American Society for Engineering Education (ASEE) Executive Committee of the Engineering Deans Council and the ASEE Corporate Roundtable suggested that with the end of the Cold War, engineering education needed a new set of guiding principles to replace those that had been developed following World War II. "Rather than a world based largely on superpower competition and national security, engineers now faced a world of intense international economic competition and widespread public uncertainty about technology and its uses."** The result was to initiate a project to prescribe changes and initiatives to bring engineering education into a posture commensurate with the new times.

A quote from Dr. Lewis Perelman's provocative book is representative of other, more urgent, views.***

> Academia's universally recognized and much lamented resistance to change cannot endure much longer. The reformation is now under way and out of control. The

---

* National Commission on Excellence in Education. *A Nation at Risk: The Imperative for Educational Reform.* Washington, DC: U.S. Government Printing Office, 1983.

** American Society for Engineering Education, "Engineering Education for a Changing World," Project Report by the Engineering Deans Council and Corporate Roundtable of the American Society for Engineering Education, October 1994, p. 8.

*** Dr. Perelman is a Senior Fellow of the Discovery Institute and formerly Director of Project Learning 2000-a study of restructuring education and training sponsored by nine U.S. corporations and foundations.

academic empire is less likely to gradually fade away than to reach a critical point at which it suddenly implodes and is overturned in a fairly short period of time.*

The seed for the Integrated Science and Technology program at James Madison University was planted in 1988 when the Virginia General Assembly created the Commission on the University of the 21st Century. The Commission's charge was to review higher education in Virginia and to recommend ways to strengthen Virginia's position among the states and ensure that Virginia would remain abreast of changing social and economic orders. In its 1989 report,** the Commission challenged each institution of higher learning in Virginia to provide leadership in bringing about needed educational reform. In response to this challenge, JMU proposed a new college for science and technology education, and in January 1993 the State Council on Higher Education in Virginia approved a new, innovative major in Integrated Science and Technology as the lead program in the new College of Integrated Science and Technology.

INTEGRATED SCIENCE AND TECHNOLOGY: THE TEACHING APPROACH

There are two broad goals for the Integrated Science and Technology (ISAT) program at James Madison University. The first goal is to provide an alternative paradigm for science education at the college level. Rather than offering a curriculum that follows the traditional approach of providing in-depth knowledge in a specific scientific discipline, the ISAT program is preparing students as generalists by providing instruction across the sciences, with emphasis on the methods and applications of science. The intent is to develop a graduate with a broad, coherent grasp of science and technology in the midst of a high degree of specialization, and one who furthermore sees science and technology in the context of "real world" constraints and requirements. Both traditional science programs and the ISAT program prepare students to solve problems involving science and technology, but the ISAT curriculum differs sharply in two ways: it is a process-oriented rather than a content-oriented curriculum, and it pays particular attention to exposing the student (as nearly as is possible) to the same real-life, professional environment he or she will encounter upon graduation.

The second hope of the ISAT program is to attract students who are traditionally under-represented in science and engineering. This includes not only minorities and women, but also many students interested in science and its applications who would not consider themselves "scientists," and would not normally enroll in a science or engineering major. Such students would like to obtain a broad understanding of science and technology, and are enthusiastic about entering a profession based on science and technology, but are unwilling or unable to pursue traditional majors in science and engineering. The U.S. has a growing need for men and women who are

---

* Perelman, Lewis J. *School's Out.* New York: Avon Books, 1993, p. 116.
** The Commission on the University of the 21st Century, "A Letter to the Governor, The General Assembly, and the People of Virginia," Richmond, VA, November 15, 1989.

conversant with technology and possess a broad comprehension of science and its methods, but who are not themselves researchers, developers, or engineers.

Apart from the curriculum, which we shall address in more detail shortly, there are seven characteristics of the ISAT program that, taken collectively, distinguish it as unique in higher education:

- **Integrated Instruction.** Instruction in the program is presented in a fashion that integrates the subjects of mathematics, science, and technology. Subjects such as calculus, physics, biology, and chemistry are not treated as stand-alone subjects. Coherence is provided largely by a context of applications, because in the real world problems almost never exist as pure problems in physics, chemistry, biology, or whatever. Real-life problems require a mixture of disciplines in their solution, and ISAT students are confronted with this troublesome truth early in their studies.
- **Team Teaching.** Integrated instruction is difficult for professors educated in a scientific or technical specialty, so most courses are developed and taught by interdisciplinary faculty teams. This approach also has an advantage in that it models problem solving in the professional world. It is rare that a single individual (professor) holds the key and all knowledge to solving a particular problem in real life. The broad spectrum of disciplines represented on the faculty, required for this teaching approach, is a strong attribute of the ISAT program.
- **Consideration of Nontechnological Issues.** A natural corollary to the fact that problems come to us in life as interdisciplinary in nature is the fact that complex, technologically based problems often have solutions constrained by politics, economics, ethics, or some other nontechnological issue. This social science dimension to science and technology is explicitly presented to the students in their freshman year, and is carried as a theme throughout the 4 years of their studies.
- **Student Collaborative Learning.** Developing students' collaborative skills is further preparation for the contemporary workplace where problems are typically solved in teams or in groups, and not by single individuals in isolation. Consequently, students are permitted, and often required, to study and solve problems collaboratively. Homework is routinely performed collaboratively by the students, and projects are regularly assigned that require the students to work in teams of four to six students. Students are graded on their teaming abilities by the professor as well as fellow team members.
- **Intrinsic Use of the Computer.** Use of the computer as a problem-solving tool permeates the curriculum. Students are given e-mail accounts and access to computers the first day of class, and are immediately required to begin interacting with their professors and peers electronically. A knowledge of simple software tools such as word processing, spreadsheets, and mathematical software are taught to students during their freshman year in computer laboratories that accompany each ISAT course. Students are also

strongly encouraged to purchase their own computers. The role of computing in the ISAT program is discussed in more detail, below, as we discuss information and knowledge management within the curriculum.

- **Use of Modern Pedagogical Methods.** Calls for reform in teaching science, mathematics, and technology emphasize improved pedagogical methods. Traditional lectures, supported by chalk and board, may be efficient ways to present material, but they have been demonstrated to be ineffective in the learning process. Student-centered teaching and inquiry-based learning methods that force students to be active in the learning process are known to improve learning and to increase student motivation and interest. In the ISAT program, these approaches are coupled with instructional technology to communicate concepts that are often foreign and arcane to the undergraduate. Furthermore, students are taught to use instructional technology in developing their *own* communication skills.

- **Development of Motivational Content.** During the freshman and sophomore years, special attention is paid to motivating students to learn. This is important for the majority of students, particularly because ISAT recruits students without strong interests in content-based instruction. Motivation comes from applications-based instruction with an inquiry-based approach. Problems are chosen with an eye to current events and topics in which students have expressed interest, such as infectious diseases (notably AIDS and Ebola), the environment, robotics and automated manufacturing, and alternative energy sources. Further, by actively engaging students in the practice of science, the faculty is better able to enable students to share in their own excitement for science.

THE ISAT CURRICULUM

The ISAT curriculum has three major parts: lower division (freshman and sophomore) foundations, junior year breadth classes, and upper division (junior and senior) classes in a concentration area.

LOWER DIVISION FOUNDATION

The foundation of the curriculum is presented through 28 semester hours of integrated instruction in the first 2 years. As we mentioned earlier, this is done through integrated instruction that includes both scientific and technological topics as well as nontechnological considerations. During these 2 years, a good deal of attention is paid to nurturing the students, and inculcating excitement for science and technology. To provide an understanding of the content of these two years, we can provide a rough equivalency in traditional terms. There is a good deal of synergy gained from the integrated instructional approach, and it is not entirely accurate to state quantitatively the content in traditional terms. That being said, however, for purposes of establishing a basic understanding of the content of the freshman and sophomore years, the traditional content is roughly as shown in Table 10.1.

TABLE 10.1
The Traditional Elements of Course Content as they Might be
Equivalent to the 28 Hours of Integrated Instruction

Traditional subject	Semester hours
Calculus	4
Laboratory science (physics, chemistry, biology)	12
Business, management, economics, related subjects	2
Personal computer applications, procedural and declarative programming	3
Statistics	3
Relevant social science courses	4
Total Semester Hours	28

JUNIOR YEAR BREADTH COURSES

During the junior year, students begin to look at science and technology in more depth and are exposed to specific areas of technology termed *strategic sectors*. The term has been chosen to describe areas of technology that have strategic significance to the nation's economy and social well being. Six broad sectors have been identified, having been distilled from national critical technologies lists and selected as those areas of technology that hold the most import for meeting our nation's technology agenda. We readily admit that six strategic sectors do not span the scope of the important national technologies, but we were forced to make choices under the constraints of staff limitations. Over time, we expect the strategic sectors to change, with more being added and some probably fading away.*

Ideally, each ISAT student would take all strategic sectors. But given a limit of 120 semester hours in our baccalaureate program, students are required to take only four of the six sectors, one of which must be Instrumentation and Measurement, a laboratory component linked to the other sectors. Currently the six strategic sectors are:

- Biotechnology
- The Environment
- Energy
- Information and Knowledge Management
- Engineering and Manufacturing
- Instrumentation and Measurement

Each strategic sector consists of 2, 3-hour courses except for Instrumentation and Measurement, which is 7 hours in length, and is composed of 1 3-hour foundational course in data collection and analysis, and 4, 1-credit lab courses associated with the other strategic sectors except Information and Knowledge Management.

---

* Communications, in particular telecommunications, will likely emerge as a strategic sector in the near future.

This approach of distinguishing distinct laboratory activities has been taken to ensure a strong hands-on laboratory experience for students during the junior year studies of technology.

UPPER DIVISION CONCENTRATION

Early in the second semester of the junior year, each student selects an area to be his or her concentration. The concentration consists of 12 hours of course work and 6 hours of senior thesis, the capstone experience of the educational program. The concentrations are, for the most part, extensions of the strategic sectors. There are two exceptions: the Instrumentation and Measurement sector does not offer a concentration, and a concentration in Health Systems has been added to take advantage of the professional opportunities for an ISAT student in health-care administration. The concentrations, then, are

- Biotechnology
- The Environment
- Energy
- Information and Knowledge Management
- Engineering and Manufacturing
- Health Systems

As we mentioned above, each concentration includes a senior thesis. The thesis is based on a student's experience as a member of a team of five or six students who work on a real-world problem that has been identified by industry or government. This provides the student with an experience opportunity that is of substantial value in bringing reality to 4 years of classroom theory and training.

INFORMATION AND KNOWLEDGE MANAGEMENT

As described above, information and knowledge management is both a strategic sector and a concentration option for ISAT students. But information and knowledge management plays a more central role in the ISAT curriculum than merely being one among several sectors and concentrations. In fact, Information and Knowledge Management has been, and continues to be, seriously considered as a required strategic sector for all students, just as Instrumentation and Measurement has been so specified due to its perceived importance.

The important role of information and knowledge management in the ISAT program is based on two observations:

- information and knowledge, and the digital technologies that manipulate them, have become the crucial factors in the economy,* and
- information and knowledge management skills and tools pervade the practice of science and technology.

---

* See Don Tapscott, *The Digital Economy*, McGraw Hill, 1996.

We will first consider each of these observations in turn, along with their consequences for the study of science and technology. Next, course work in information and knowledge management in the ISAT program will be described, and, finally, we will discuss the role of information and knowledge management as an integrating discipline throughout the curriculum.

INFORMATION AND KNOWLEDGE AS KEY ECONOMIC FACTORS

The observation that information and knowledge have become the crucial economic factors is based on the shift the global economy has been undergoing since mid-century from an industrial to an information-based economy.* Land, labor, and capital were all of fundamental importance for the industrial revolution, but their relative importance has waned in the information age. Highly developed extraction and transportation technologies, combined with competitive global markets, have made raw materials abundant and cheap; labor is more expensive today, but extensive automation has made it less necessary than during the industrial revolution; capital is still a vital resource in manufacturing and transportation, but modern first-world economies are based on service and information industries. Meanwhile, the importance of information and knowledge have increased as automation and computerization have been incorporated in every aspect of business, government, and private life. Information and knowledge are the key factors in the emerging Information Economy.

Realization of the fundamental economic importance of information and knowledge suggest that they should be studied and managed as closely and carefully as land, labor, and capital traditionally have been; information and knowledge should be treated as key corporate assets.** The emerging discipline of *information and knowledge management* studies systems, processes, and techniques for creating, acquiring, organizing, controlling, and using information and knowledge.***

The central role of information and knowledge in the global economy, and hence in the plans, policies, and activities of every modern enterprise, especially those involved in science and technology, requires familiarity with the basics of information and knowledge management. A corollary of this requirement is that every student of science and technology should be well-grounded in this discipline.

PERVASIVENESS OF INFORMATION AND KNOWLEDGE MANAGEMENT SKILLS AND TOOLS

The computer has become an indispensable tool for virtually every knowledge worker in society, and particularly for scientists and technologists. Besides the now mundane uses of computers for word processing, data storage and retrieval, and electronic mail, scientists and technologists are increasingly using computers for in-depth and thorough data analysis and display, modeling, and simulation as a faster and safer way to investigate phenomena, internetworking for publication and

---

* See Peter F. Drucker, *Post-Capitalist Society,* Harper Business, 1993.
** See Peter M. Senge, *The Fifth Discipline,* Doubleday Currency, 1990.
*** Karl M. Wiig, *Knowledge Management Foundations,* Schema Press, 1993.

dissemination of research findings, and so forth. Some have suggested that recent advances in modeling and simulation, and in data visualization, are fundamentally and radically changing the scientific method itself, and that we are seeing the emergence of new paradigms for investigation.

Another area where information and knowledge management approaches have become standard is in the study, description, analysis, and improvement of processes and systems in business, industry, government, education, and so on. Methods once only known to systems and industrial engineers and computer scientists have now become standard fare in discussions of total quality management, business process reengineering, enterprise restructuring, enterprise data modeling, and the learning organization.

A final area where information and knowledge management techniques are becoming important is in building smart systems. Humans have built both knowledge and intelligence into their environment for millennia. For example, the use of templates in construction and manufacturing is a way to build the knowledge of a master into a tool that can be used by the apprentice, and using thermostatic temperature control is a way to build intelligence into a heating or cooling device. But recent advances in software development expertise, support tools like expert systems and neural networks, and the increasing ubiquity of computers, have made it possible to build more and more knowledge and intelligence into the human environment. Our environment is thus able to shoulder increasingly greater parts of the burdens of everyday existence; we are rapidly approaching a time when our environment will take care of us more than we take care of it.

Expertise with information and knowledge tools that support the use and development of science and technology, the ability to study, describe, analyze, and improve processes and systems, and the ability to modify the environment to increase its knowledge and intelligence are in some degree essential for every scientist and technologist. Thus every student of science and technology should acquire understanding and skills in information and knowledge management.

Information and knowledge management is both a strategic sector and a concentration in the ISAT program. In addition to these roles, however, information and knowledge management material is present throughout the ISAT curriculum, and as such it plays an integrative role in the curriculum. The remaining sections of this chapter consider these roles of information and knowledge management in the ISAT program.

INFORMATION AND KNOWLEDGE MANAGEMENT SECTOR AND CONCENTRATION

Students who elect to take the Information and Knowledge Management sector (consisting of two courses in the junior year, as explained above) are given a broad introduction to knowledge management, and practice in applying information and knowledge management methods and techniques to support problem solving in other areas of science and technology. The first course in the sector is an information and knowledge systems project course. Students form teams of three or four and build or enhance an information or knowledge management system for a client on or off campus. The systems built or modified must support work in some other area of science, technology, or business outside the computer and information fields. The course both draws on and is expected to support work in the other strategic sectors, such as The Environment, or Engineering and Manufacturing. This course hones students' problem-solving and implementation skills, and helps students see how information and knowledge processes and systems support the other sciences, business, and government. Finally, the course continues to develop in students the integrated problem-solving perspective characteristic of the ISAT program and essential to the information and knowledge management workplace.

The second course in the information and knowledge management sector sequence is a detailed overview of the field. The course covers the following:

- Definitions of data, information, knowledge, process, and system, their fundamental characteristics and relationships, and their measurement, management, and control.
- The data, information, and knowledge life-cycle, the systems that support it, and how the process and its supporting systems have evolved through history and continue to evolve today.
- Basic information technologies, their evolution, their future, and how information technologies have interacted with processes, systems, and societies today and in the past.
- Moral, ethical, legal, and organizational issues surrounding the collection, storage, retrieval and use of data, information, and knowledge, such as intellectual property, privacy and security, censorship and access to information, etc.
- An introduction to process and system modeling and analysis, including the nature of models, the basic modeling and analysis method, and basic modeling and analysis tools such as flow diagrams, linear programming, simulation, etc.

Students who have finished the information and knowledge management sector will have been exposed to all major topics in the area. Students who elect to study this area in greater depth by concentrating in information and knowledge management will take four additional courses and do a thesis in the area. The four IKM concentration courses are

- An introduction to the software industry that focuses on the software life cycle, software project management, and current trends in the industry. Students work in teams to carry a product through the entire life cycle, producing work products like a project plan, a software design, a program, a test plan, and user documentation.
- An in-depth course in intelligent systems. This course covers object-based (frame-based) expert systems, neural networks, hybrid intelligent systems, and intelligent system development strategies, particularly knowledge acquisition and representation techniques.

- A course in the multimedia industry. Students learn basic design principles and techniques and a multimedia production environment, and produce at least one significant multimedia product. Students also learn about trends and emerging technologies.
- An additional course in information and knowledge management drawn from offerings in computing, data processing, telecommunications, artificial intelligence, and so on, offered by other departments on campus.

The concentration courses provide in-depth coverage of several topics for ISAT majors concentrating in information and knowledge management.

INFORMATION AND KNOWLEDGE MANAGEMENT AS AN INTEGRATING DISCIPLINE

As noted above, information and knowledge management tools pervade contemporary practice in science and technology. Thus, one would expect to find information and knowledge management methods, tools, and techniques taught and used throughout curricula in science and technology. Furthermore, the pervasiveness of this material helps lend coherence to a program in integrated science and technology: students will encounter the same tools, the same problem-solving approaches, the same simulation and modeling techniques, the same research techniques, and the same dissemination paths from class to class and project to project. Indeed, this is true of the ISAT program.

The ISAT lower division foundation covers the basic computer productivity tools whose use is required of any knowledge worker, as well as basic tools for modifying the computational infrastructure to build more knowledge and intelligence into the environment (in other words, computer programming). Networking and productivity tools are introduced in the first freshman course and used throughout the ISAT program, and computer programming occupies the bulk of the last course in the foundation sequence. Included among the networking and productivity tools that the students learn to use are electronic mail and the Internet (Netscape, ftp, e-mail, etc.), the Microsoft Office suite, Microsoft Project, and Jump (a statistical analysis and modeling tool). Students learn procedural programming in Visual Basic, and declarative programming in a rule-based expert system shell with frames, such as EXSYS. Programming is taught in the context of a development life cycle, and grounded in propositional logic and process modeling.

Later courses in the sector and the concentration use all of these tools, and generally add more. Particularly prevalent is the use of simulation and modeling tools throughout the ISAT curriculum, some examples of which are the following:

- Students in the Engineering and Manufacturing sector learn to use Sim-Factory to simulate an automated factory and to solve manufacturing floor layout and supply problems.

- Students in the Biotechnology sector or concentration may take a course in protein modeling in which they research a protein on the Internet, and create a World Wide Web home page about the protein that explains it and models it using a three-dimensional protein modeling program.
- Students in the Environment concentration may take a course in environmental simulation and modeling where they create computer models of some aspect of the environment as the key tool in solving environmental problems.
- All students must prepare both written and oral results as their thesis. Oral presentations will typically use multimedia presentation tools such as PowerPoint, Astound, or Netscape, and written results will typically be in the form of World Wide Web pages.

Information and knowledge management is essentially a support technology, and in this role it is able to add connection and continuity to students' experiences in the ISAT program.

The intent of integrated instruction in science and technology is to develop problem solvers able to consider and appreciate the full range of factors affecting a science or technology problem, to apply orderly problem-solving techniques, to generate innovative solutions satisfying all needs and constraints, and to be able to document, communicate, and if necessary implement problem solutions for clients. We might call such a problem solver a *systems thinker* to emphasize that the aspects of the entire system surrounding a problem are taken into account and dealt with by the problem solver. Developing such a systems thinker is precisely the intent of the ISAT curriculum.

One discipline stands out as incorporating, as part of its fundamental subject matter, the unifying concepts, techniques, and methods that characterize systems thinking: information and knowledge management. Thus, the most fundamental way that information and knowledge management serves to integrate a program in science and technology is by providing a focus and a center for the systematic study of the basic concepts, tools, techniques, and methods of systems thinking. Information and knowledge management is thus the methodological foundation of ISAT, and as such provides the unified basis that integrates study throughout the curriculum.

In summary, information and knowledge management plays an integrative role in science and technology education because, first, its tools, methods, and techniques so pervade all study of sciences and technology, and second, because information and knowledge management provides the methodological foundations of the systems thinking that distinguishes the interdisciplinary problem solver in integrated science and technology.

CONCLUSION

Information and knowledge management is an essential component of an innovative program in science and technology education at James Madison University. Although the program is still young, it is clear that fundamental information and knowledge management methods and techniques ground Integrated Science and Technology majors as systems thinkers and problem solvers, and that the information and knowledge management tools are among the most important and widely used by students as solvers of contextual problems in science and technology. Furthermore, the material knowledge of information and knowledge management, and the skills associated with it, prepare ISAT students to participate fully in the Information Economy.

REFERENCES

1. American Association for the Advancement of Science. *Science for All Americans.* New York: Oxford University Press, 1990.
2. American Society for Engineering Education. *Engineering Education for a Changing World.* Washington: American Society for Engineering Education, 1994.
3. Clark, Mary and Wawrytko, Sandra. *Rethinking the Curriculum: Toward an Integrated, Interdisciplinary College Education.* New York: Greenwood Press, 1990.
4. Drucker, Peter F. *Post-Capitalist Society.* New York: Harper Business, 1993.
5. Perelman, Lewis J. *School's Out.* New York: Avon Books, 1993.
6. Senge, Peter M. *The Fifth Discipline.* New York: Doubleday Currency, 1990.
7. Tapscott, Don. *The Digital Economy.* New York: McGraw Hill, 1996.
8. Tobias, Sheila. *Revitalizing Undergraduate Science: Why Some Things Work and Most Don't.* Tucson: Research Corporation, 1992.
9. Wiig, Karl M. *Knowledge Management Foundations.* Arlington, TX: Schema Press, 1993.

11 Educating Knowledge Engineering Professionals

Johan C.M. den Biggelaar

CONTENTS

INTRODUCTION

This chapter is about a nonconformist approach to educating knowledge engineering professionals in an institute for higher education. To understand the specific view on the education of Knowledge Engineering professionals as presented in this chapter, some background on the environment is essential.

    The main mission of Kenniscentrum CIBIT is to enable industry to compete better in a global economy and to enable nonprofit organizations to use their

169

resources better for the improvement of service standards through the effective application of innovative Information Technology.

Before the Center was founded in 1988, quite some time was spent on choosing the most effective strategy for the required technology transfer. It was found that joint research projects can be useful in the technology transfer between universities and -big- companies in a (Research and Development) R&D setting, but not for direct operational application of the technology.

It was decided that an advanced training of software engineers would be a much more direct and effective strategy. Therefore, many of the available resources were spent on developing a professional Master's course in Knowledge Engineering. Originally, the traditional academic approach to course development was "We teach what we find interesting and/or important."

After quite some energy had been put into the development of the course structure and course materials, the design was discussed with the business community. As the Center had to generate its own income from the services provided after the initial start-up period, it was essential to listen to the market carefully.

From the discussions with industry, we learned that the problem of course requirements had to be approached from a different perspective: the business requirements of knowledge engineers. We had fallen into the pitfall of traditional academic product-orientation and had produced a course syllabus that was rather traditionally AI-oriented. In fact, it was the kind of AI-syllabus that can still be found in universities in many countries.

Education and training in any professional field must have strong links with existing professional practice in industry. In academic courses, one would expect an emphasis on current scientific issues. The truly professional course must provide a balanced mix of the best of both worlds. Students must be trained in applying professional practices but should also be able to introduce the results of recent scientific research in their workplaces and act as "agents of innovation."

This chapter is based on practical experiences in the training of knowledge engineering professionals for the Dutch labor market. Since we know from international research that our economy is in some ways different from several other countries, some particularities will be described in a few paragraphs below for the necessary context.

The core of our experience is based on a professional Master's course for knowledge engineering professionals. Since 1988, over 200 students have participated in this 2-year part-time course. In the course of the last 2 years, we have gained extensive experience with short courses in industry as well. These courses are often taught in-company. Most clients have established knowledge engineering groups developing in-house knowledge-based systems or they provide knowledge engineering services to other companies.

Professionals in the field in the Netherlands seem to have reached a consensus on the need for a broader concept than knowledge engineering per se. Knowledge Technology is now an established term and an increasing number of professionals in the field put the title "kennistechnoloog" ("knowledge technologist") on their business cards. The same development is currently taking place in the Scandinavian countries.

A similar development may be observed in the area of systems terminology. Terms such as AI and Expert Systems are mostly avoided for known reasons. Recently, the term "Knowledge-based System" is going the same direction, since it is mostly used in conjunction with stand alone systems. The new term which is gaining recognition is "knowledge-intensive system" which refers to the new generation of totally integrated information systems with knowledge-based components.

KNOWLEDGE TECHNOLOGY IN THE NETHERLANDS

Since the Dutch economy has a strong position in financial and other services internationally and is relatively weak in industrial production, this is also reflected in the IT market. The Dutch IT industry is mostly a service industry. Since the Dutch market is relatively small, only a handful of companies have developed and marketed standard software packages. Internationally known are Uniface and Baan's Triton, a serious competitor for SAP's R/3.

As a result, IT companies hardly invest any efforts in R&D. Much effort is spent on developing standard methodologies and working practices to be able to provide efficient services in a strongly competitive market. A relatively small number of IT professionals are working in product development. Most are working in software services "body shopping" companies or in IS departments of big service and government organizations.

Since there is so much emphasis on software services and third party development, it is no coincidence that the KADS research was started at the University of Amsterdam. The result, CommonKADS, has become an almost de facto standard methodology in Europe for developing knowledge-intensive systems. For some years now, most companies that provide commercial services in this area have adopted many elements of CommonKADS in their general methodology for developing information systems.

With the continuing integration of IT and knowledge technology, however, the need for one conceptual framework for modeling information and knowledge has grown. Recent work at our center in using the Yourdon modeling approach for both data processes and knowledge has resulted in SDF, System Development Framework, which is now being widely used in industry. (Van der Spek, 1993; Kruizinga, 1995)

Our approach has been successfully applied in many projects in industry. Since this approach can easily be supported with existing CASE tools, knowledge engineering is becoming more and more integrated in standard software engineering practices. This way, knowledge engineering has become an extension of the capabilities of well-trained software engineers and the need for dedicated knowledge engineers as a new species is diminishing.

This development is amplified by the developments in current tools for the development of information systems. The "traditional" expert system shell was replaced by more general development tools such as AionDS and — for real time systems — G2. More and more features of these tools can now be found in more traditional IT-tools such as Oracle and other database-oriented development tools. As a result, the distinction between the world of knowledge engineering and information technology is diminishing too.

Well, you may argue, if this is all true, what remains of the field of knowledge engineering? To start with, there is no simple answer to that question but two trends can clearly be seen:

KNOWLEDGE MANAGEMENT

In Management Science, a strong interest in the economic value of expertise for organizations has developed over the past 3–4 years. This development has been boosted by the experiences in Business Process Redesign projects in the services industry where modern IT concepts such as client server and knowledge-based systems have proven to be important tools for implementation of the new IT strategies.

Moreover, the field of knowledge engineering has developed the concepts, methods, and techniques, such as *knowledge modeling*, to support management consultants into turning their abstract ideas in more effective application of knowledge into practical solutions.

Knowledge engineers with an additional background in management science will be able to apply their skills in this field over the next few years as consultants in Knowledge Management and Knowledge Modeling (Wiig, 1993).

SOFTWARE ENGINEERS "PLUS"

Software engineers with an additional thorough training in knowledge engineering have the right skills to build flexible, maintainable systems using modern high productivity tools based on object technology in a short time. There is a growing need for such software engineers "Plus" with a service attitude and excellent communication skills who take the user seriously.

Their expertise is also very valuable in the area of datamining and cbr-applications.

Unfortunately, industry in the Netherlands has no need for academically trained researchers and other specialists in "cognitive informatics" or "Artificial Intelligence." Most graduates from such programs have a hard time finding jobs. Skills in Lisp or Prolog programming or even neural networks are mostly useless in today's IT industry and can only be applied within the walls of academic institutions.

The skills that industry needs are clearly going in a totally different direction.

RETHINKING THE TRAINING OF KNOWLEDGE ENGINEERING PROFESSIONALS

From regular contacts and formal interviews with hundreds of professional knowledge engineers, software engineers, and their managers we derived a so-called "skills profile" of a knowledge engineering professional. This is a superset of the set as described by Jay Liebowitz in 1991 (Liebowitz, 1991).

Our main additions are in:

- More extensive software engineering skills
- A better understanding of current organizational issues, including knowledge management

These are mainly the results of our observations of the advances in professional practices in Knowledge Engineering during the last 3 years. An informal and rough representation of the set we currently use is given below:

1. Interpersonal communication with:
   - Domain experts
   - End-users
   - Clients, management
   - Software engineers, system engineers
   - Other project team members
2. Business and organization:
   - Domain knowledge
   - Understanding of current and potential applications in the domain environment
   - Understanding of organizational issues, including business process redesign
   - Skills in technology transfer
   - Skills in project management techniques, both traditional as well as risk-driven
   - Understanding of general business operating environments
   - Understanding the basics and value of knowledge management
3. Analysis and design skills:
   - Knowledge acquisition: techniques for knowledge analysis and knowledge elicitation
   - Excellent skills in knowledge modeling, user modeling, interaction modeling and organization modeling; "traditional" as well as object oriented
   - Designing modular systems using CASE tools, which are easily extendible and maintainable
   - Evaluation and selection of the proper tools for the task at hand
   - A strong theoretical and practical background in software engineering
4. AI-technology skills:
   - Broad understanding of AI-fundamentals, including logic and knowledge representation formalisms
   - Broad knowledge of the tools, including CASE tools, rule-based tools, and object-based tools
   - Understand what can and cannot be achieved with current technology
   - Understand information technology theory and practice
   - Software engineering principles and methodology
   - How to integrate knowledge-intensive system components with "conventional" information systems, databases
   - Understanding of man-machine interfacing techniques, including GUIs

It is no coincidence that the interpersonal communication skills are first on the list. We find these very important, since the success or failure of most projects depends on these skills. During the course, quite some time is spent on training these interpersonal skills, not as a separate subject, but as an on-going activity from the start to the end of the course. This longitudinal approach is necessary to develop the right attitude with students. It also requires that special attention be paid to the selection of candidates for the course. A brilliant science student who is socially retarded may be an excellent researcher or programmer but will be quite inadequate as a knowledge engineering professional.

We have reduced the number of lectures and replaced them with workshops and case studies which are solved in teams of 2–6 students.

We must realize that for several years to come, the number of practicing knowledge engineering professionals will still be relatively limited. Therefore, they will often be pioneers in their organizations. Since the success of the technology transfer process to their organization depends to a great extent on their personal skills, they need to be trained to manage this process and to act effectively in "selling" the technology.

THE MASTER OF SCIENCE COURSE IN
INFORMATION AND KNOWLEDGE TECHNOLOGY

The course aims to produce "Masters" in the building of knowledge-intensive systems. This goal is reached by providing students with a thorough theoretical background and a variety of skills, as described above. The course accepts IT-professionals with at least a bachelors degree in Computing Science or Business Information Systems and at least 3 years practice in industry.

Considerable emphasis is put on methods and techniques for knowledge acquisition and modeling. Graduates of the course have sufficient technical and methodological knowledge to judge whether or not a specific task can be formalized in a knowledge-intensive system. Moreover, graduates of the course are trained to implement such systems.

The objectives of the course are to provide the students with:

- the right skills to perform in practice as professional engineers with the profile as described above,
- a good insight in the direction of current developments in Knowledge Engineering, and
- familiarity with important sources of information such as international journals, conferences as well as organizations in the field

The course has a modular structure. The academic year lasts from the beginning of September to the end of June, which is about 10 months. The course is part-time, lasts 2 years and starts twice a year. Four subjects of 2.5 months each are taught in the first year and two in the second year. The final project takes about 6 months. The course starts twice each year, in September and in February. Almost all students are being sponsored by their employers.

The course is not part of any university-wide modular scheme. The subjects are only part of this particular course, which results in a well-integrated program. Through the modular design, the students, who mostly have full-time jobs in industry, can concentrate on one set of related subjects, for example, learning how to do knowledge acquisition and modeling, which is tested at the conclusion of that subject.

Due to the part-time character of the course, students spend only an average of 1 day a week at the institute. Therefore, we have developed hundreds of pages of course materials for every subject in order to make the learning process as effective as possible. As part of their course work, they use professional rule-based and object-oriented tools such as AionDS and G2 at their home PCs.

Starting with the subject "Modeling of Knowledge-Intensive Information Systems," the course uses several case studies as a vehicle for carrying out assignments.

The teaching method focuses on the development of good engineering practice. The lecture material directly relates to, and provides thorough theoretical fundamentals for the project activity.

Students are supervised intensively by experienced tutors. No more than 18 students are accepted in every cohort. Practical exercises are corrected by the academic staff to ensure a maximum feedback between knowledge transfer and conception. The progress of all students is monitored through regular meetings of the course team. The objective of the course team is to make the course demanding as well as enjoyable and stimulating.

SOME EXPERIENCES WITH 15 COHORTS OF MASTER'S STUDENTS

Since the start of the course in September 1988, 15 cohorts of students have enrolled. At present the course has about 70 students divided over 4 groups.

The first few cohorts consisted of students with a wide range of backgrounds, from mathematics to social sciences. This situation has changed considerably, however. The last 8–9 cohorts consisted mainly of experienced software engineers. Most of these students already had a degree in computer science or in information systems. The others had in general a combination of a formal education in a different discipline with additional industrial training in IT.

The course has the character of an "executive" engineering course. The youngest student is 25 years old, while the average age is over 30. The students seem to opt for a further career in engineering (management) rather than general management and wish to renew and extend their knowledge and widen their scope. They are extremely well motivated and are eager to learn.

Moreover, since they are mostly pioneers in their own organizations, they wish to stay in contact with each other after graduation. Therefore, we have founded an association for present students and alumnae. This is becoming a strong "old-boys-network" with the Center as its home base. Many students have already made a career in their organizations and are now knowledge engineering group managers.

INTERNATIONAL LINKS

The course team has established a cooperation with the faculty of Technology at Middlesex University in London, Great Britain since 1989. In that year, the course was validated jointly as a British Master of Science course in Knowledge Engineering under supervision of the British Council on National Academic Awards, the

CNAA. Middlesex University has started a similar course in September 1991 with a more traditional AI profile, but has adopted our course design and contents with a new group of staff in 1995. Graduates from Utrecht receive two degrees: a postgraduate diploma from Utrecht and, since 1990, the Master of Science degree in Knowledge Technology from Middlesex University.

From many international contacts, we have learned that the current course design is rather unique in its strong focus on knowledge engineering as a professional discipline. Several other universities in Spain and other European countries have adopted parts of our approach.

MODULES OF THE CURRENT COURSE

The course's content is reviewed and adjusted annually. Subjects such as Lisp and logic programming have been removed years ago. (Biggelaar, 1992).

The current Master's course has six modules and a final project in industry. Material from all of these modules is frequently repackaged for use in industrial in-company training programs. Therefore, the course as a whole is a good representation of what is needed for state-of-the art training in a modern knowledge engineering curriculum. Below are given descriptions of the various modules in the course:

INTRODUCTION IN INFORMATION AND KNOWLEDGE TECHNOLOGY

This module offers insight into the role of information technology (IT) in organizations and in modern approaches to the concepts "information" and "knowledge." The module also provides a thorough introduction into model-based development of knowledge-intensive information systems. Discussion will focus on:

- IT's role as "enabler" in the framework of the redesign of business processes (BPR)
- The role of knowledge in organizations and an effective management of knowledge: knowledge management
- Modern approaches to Project Management, including risk-driven PM and RAD/IAD
- Basic aspects of information systems: objects, processes, and control
- The added value of knowledge-oriented system development
- The role of object orientation
- Principles of software engineering and modern architecture, including client server
- The System Development Framework
- Application areas and characteristics of knowledge-intensive systems

MODELING OF KNOWLEDGE-INTENSIVE INFORMATION SYSTEMS

This module offers a detailed discussion of the framework for the model-based development of knowledge-intensive information systems, SDF. The student becomes skilled in a structured approach to system development, based on models and scenarios. The student not only gains insight into the methods and techniques, but also acquires practical experience in the modeling of basic aspects of information systems. Attention is paid to the modeling of knowledge in the framework of system development.

The following subjects will be discussed:

- A framework for the structured development of knowledge-intensive information systems
- The determination of scenarios as the basis for functionality ("event partitioning")
- The modeling of basic aspects: objects, processes, and control
- The relationship between knowledge and other basic aspects
- Specific aspects of CommonKADS and the application of templates for Knowledge Modeling
- Techniques for the modeling of knowledge, including OMT (Rumbaugh, 1991)

The module is concluded with a workshop in which a case will be worked out with the help of an industry standard CASE-tool, which was adapted and extended to support knowledge modeling.

The Center has decided to license the System Development Framework (SDF) to industry. It is now being used by several companies for in-house use as well as in service-based projects.

MODELING OF IMPLEMENTATION ASPECTS

The student develops skills in detailing the basic design of a knowledge-intensive information system. The student makes use of object-oriented design principles, based on Object Modeling Technique.

Further, the student develops skills in the modeling of human-computer interaction (HCI) and in the integration of various system components (databases, data bank, applications, and user interfaces).

Subjects include the following.

- Modern software engineering principles
- Application of object orientation in the design phase (OMT)
- Design of rule collections and reasoning strategies; tools for quality control of knowledge bases
- Practical guidelines for the determination of human-computer boundaries
- Determination of the software configuration and technical infrastructure
- Aspects of human-computer interaction
- Detailing of system components

During the concluding workshop, the student designs the implementation model of a knowledge-intensive information system.

Techniques for the Analysis of Knowledge-Intensive Processes

This module discusses the techniques that can be used to analyze knowledge-intensive tasks. This is generally called "knowledge acquisition." Examples of the techniques discussed are interviewing, thinking aloud, card sorting, and simulation techniques. The teaching methods in this module aim at the development of practical skills in the application of these techniques. The module also examines the characteristics and the possibilities for the application of "adaptive software" (case-based reasoning, neural networks, and genetic algorithms) as an aid during the process of analysis and for data-mining purposes.

To sum up, the module discusses the following subjects.

- The development of an analysis plan
- Techniques for gathering knowledge from domain experts, mainly by practical training with real domain experts from specialized domains, such as a music recording technician and a medical expert
- Effective communication with domain experts
- Techniques for knowledge acquisition from text sources
- Principles of "learning software" and adaptive systems, including neural networks, genetic algorithms, case-based reasoning, etc. for data-mining

The module concludes with an extensive workshop, during which the acquired skills are applied in an integrated manner with the help of a case.

Knowledge and Information Systems in Organizations

To an increasing extent, knowledge is seen as the crucial production factor in businesses and other organizations. Accordingly, interest in the control and management of critical knowledge areas is growing quickly. The use of knowledge-intensive systems is one of the tools to optimize the knowledge household.

In this module, the student gains insight into the organizational aspects of the development and introduction of knowledge-intensive information systems in organizations. The module also pays a great deal of attention to project management aspects such phasing, planning, risk analysis, and quality control. Both a standard linear "waterfall" approach and modern risk-driven approaches are discussed.

Among other things, the module discusses:

- Principles of Business Process Redesign (BPR)
- Strategic policy making concerning knowledge in organizations
- Concepts, strategies, working methods, and techniques for knowledge management
- Application of knowledge-intensive information systems
- Cost benefit analysis of knowledge-intensive information systems
- Quality of systems

- Risk-analysis of projects
- Risk-driven project management and Rapid Application Development (RAD)
- User-oriented development
- Organization of maintenance of knowledge-based and information systems.

REALIZATION OF SYSTEMS

This module discusses the realization of knowledge-intensive information systems. The student is familiarized with advanced tools and development environments for storing data and knowledge. The module discusses the following subjects:

- Characteristics of technologies which play a part in the realization of knowledge-intensive information systems
- Standards for communication, storage, and distribution (especially OO and client server concepts)
- Selection of hard- and software components
- Characteristics of distributed systems
- User-interface technology, among other things, graphical user interface design (GUIs)
- Testing, validation, and maintenance of systems.

All aspects of the realization of systems are discussed in an extensive workshop, aimed at the development of practical skills.

FINAL PROJECT

After the teaching part, every student conducts a final project in industry, which has a student load of about 500 hours. During this project, the student first has to develop a Project Plan as the basis for the project. During the project, he or she is supported by a member of staff of the Center and a coach in the company where the project is done (not necessarily his or her own company).

The final examination has three parts: the written thesis, the oral presentation and defense of the thesis, and a visual presentation of the project results in the form of a poster.

CURRENT AND FUTURE DEVELOPMENTS

Currently the program is undergoing one of its regular revisions. The course is gradually developing towards the education of IT engineers who have a specialized toolbox to support the active management of knowledge assets of an organization. Since traditional knowledge engineering provides only a limited set of solutions to problems in this area, we are currently developing two alternative routes in the program, one of which is totally integrated while the other leads to an extra optional module.

The first development which has been taught as different bits and pieces in several modules is now becoming a major stream in the course with implications in all modules. In our point of view, Knowledge Discovery is an important new field of development and added value to be claimed by knowledge technology professionals.

The second development is the use of groupware for knowledge sharing. This may take the form of intranets or specialized groupware packages, such as Lotus Notes. The focus here is not on the tools, but on the concepts and architecture of a knowledge management infrastructure, which supports the effective sharing and use of less formalized knowledge in organizations.

CONCLUSIONS

Our strategy to use an advanced course for technology transfer has proved to be effective. Many of our graduates play a key role in the application of knowledge technology in their own companies. Moreover, the constraint of using a marketing approach has forced us to listen carefully to the requirements of the market. This has resulted in a different approach to the course design and has lowered the threshold for experienced professionals to "return to school" for a major upgrade in their knowledge and skills.

The success of the course also shows that it is possible to use a marketing approach for technology transfer, which is actually a "technology push," provided you have the right product for the right group at the right moment. The marketing lessons we have learned from the course have provided us with a completely new approach to course conception, design, implementation, and promotion.

As a result, the courses at the Center show little resemblance with traditional academic courses in AI, Knowledge-based Systems, or Expert Systems.

FUTURE DIRECTIONS

The Center for Knowledge Technology was founded in 1987 as a cooperative venture between academic institutions, the government, and the business community. It was designated as the National Center of Excellence in knowledge-intensive systems in higher professional education. The Center's objective is to develop expertise in the area of KBS-development and to transfer this expertise to individuals and organizations.

Recently, the Center for Innovation of Business Processes with Information Technology ("Kenniscentrum CIBIT") was formed as the result of a merger between the Center for Knowledge Technology and another center of excellence in Utrecht Polytechnic, "KenTel," which runs a similar Master's course for professionals in telecommunications, networks, and information systems called "TeleMatics."

The newly formed center will widen the scope of activities, providing both organizational and IT support to an effective and efficient "knowledge infrastructure" which will remain the main focus of its activities.

ACKNOWLEDGMENTS

The design and implementation of the course has been an intense team effort. We have spent several man-years on the development of course materials, which of course is an immense on-going effort of the current course team. Most of the effort to design and develop the original course has been put in by colleagues Cor Baars, Eelco Kruizinga, and Rob van der Spek.

REFERENCES

1. Biggelaar, An integrated Master's course in Knowledge Engineering, Proceedings of the Third Symposium of the International Association of Knowledge Engineers, IAKE, Washington 1992.
2. Hammer, M. and Champy, J., *Reengineering the Corporation*, Nicholas Brealey Publishing, London, 1993.
3. Liebowitz, J., Education of knowledge engineers, *Heuristics,* volume 4, number 2, Summer 1991.
4. Kruizinga, System development framework for knowledge intensive information systems, *ITESM*, Mexico, 1995.
5. Schreiber, *KADS A Principled Approach to Knowledge-based System Development,* Schreiber, Wielinga, and Breuker, Eds., Academic Press, New York, 1993.
6. Van der Spek, R. and De Hoog, R., Towards a methodology for knowledge management, *ISMICK,* Compiegne, J-P.A. Barthès, Ed., IIIA (Institut International pour l'Intelligence Artificielle), 1994.
7. Wiig, *Knowledge Management Foundations*, Schema Press, Arlington Texas, 1993. *A Knowledge Management Framework. Practical Approaches to Managing Knowledge,* Schema Press, Arlington, Texas, 1994.

12 The Other Side of Knowledge: Avoiding Descartes' Error When Managing Knowledge

Jack Presbury, Joe Marchal, and Jerry Benson

CONTENTS

> "Reason respects the differences, and imagination the similitudes of things"
> Percy Shelly, 1821 (in Gelernter, 1994, p. 80)

INTRODUCTION

A good deal of attention has been paid to developing *sound* Knowledge-Based Systems (KBS). In theory and in practice, we know how to engineer a knowledge-base that *contains only domain expertise*, build an inference engine that *generates only expert advice*, put them together in a KBS, and verify and validate the resulting product. However, we are less knowledgeable and practiced in delivering *valuable* KBS, that is, *systems guaranteed to be used and to increase productivity*. This discussion will focus on explaining why users may fail to find delivered KBS valuable. The explanation offered turns on the key role of the *intuitive-emotional* side — the "other side of knowledge" — in knowledge management. The discussion will (1) identify and analyze a serious, potential misunderstanding of the role of the

0-8493-3116-1/97/$0.00+$.50

intuitive-emotional side of knowledge in KBS theory inherited from its rationalist progenitors; (2) review alternative analyses of the intuitive-emotional side of knowledge that promise a better understanding of its role in knowledge management; and, (3) offer an explication of this other side of knowledge that, combined with the rationalist critical-rational analysis, better supports the development and delivery of KBS that are both sound and valuable.

The engineering-manufacturing metaphor of knowledge-based (or expert) systems development has knowledge first being "mined" from the data banks of experts in a particular domain, then represented and engineered to work with an inference engine that will deliver its results through a friendly interface to the user. Knowledge in this sense is a commodity; first as raw material, then as finished product. The knowledge engineer skillfully elicits (mines) the raw knowledge, then just as skillfully represents and engineers it as product, e.g., as rules or frames, for inclusion in a KBS. When the manufactured product has been checked for soundness and judged ready for market, the knowledge is conveyed through a delivery system to a consumer who judges it to be of value, and who profits from its use. This account focuses on engineering the knowledge as a sound critical-rational product and rests on a rationalist critical-rational analysis of knowledge. But there is "the other side of knowledge," the intuitive-emotional side. This has to do with the experience of the person to whom the product is being delivered.

There is a dual sense in which we are focusing on the other side of the KBS equation: first, we are focusing on the user; and second, we are focusing on the side of knowledge that motivates use of a KBS, viz., the intuitive-emotional side. This must be of concern to knowledge engineers, because without a good understanding of what motivates users, our commodity goes unsold. Knowledge in a KBS is never really valuable until a user deems it so. Until then, our carefully crafted commodity is merely potential information — or worse — meaningless noise.

On the side of knowledge that represents our logical reasoning process, there are clear KBS examples where the computer has outstripped human abilities, and because of this, it can help human users improve these reasoning processes. But the other side to human thinking, the intuitive-emotional side, is virtually ignored by KBS technology, and it is this side that we must begin to understand and to model. What makes a user receptive to knowledge in KBS? How do we engineer KBS to fit a user's critical-rational *and* intuitive-emotional schema, i.e., how do we engineer KBS to fit the way users really think? *Only through a synthesis of both sides will users experience KBS as both sound and valuable.*

DESCARTES' ERROR (VIRUS) AND THE RATIONALIST ROOTS OF KBS

Descartes' philosophical program epitomizes the Rationalist roots of the type of symbol processing approach exemplified by Newell and Simon (1990/1976). In particular, the theory and development of knowledge-based systems (KBS) has interesting parallels and shares serious problems with the Cartesian program to ground knowledge in certainty. First, both programs require a secure starting point for the knowledge in which they are interested. In the case of KBS, the initial knowledge-base is justified by containing only *certified domain expertise*, that is, expert certified beliefs about the content and procedures of some of the domain. In Descartes' case, the start-up examples of knowledge are justified by being examples that are *epistemologically certain*, that is, presented to his mind so clearly and distinctly that he could have no occasion to doubt them (Descartes, 1637). Second, once their start-up examples of knowledge are justified, both programs generate and justify adding knowledge to a knowledge base by using reasoning techniques that pass on the original guarantees. In the case of KBS, all of the system's outputs, be they descriptions of the knowledge-base, results of its reasoning, or explanations of the system's behavior, are epistemologically justified by their relation (via the inference engine) to the critical-rational status of the original certified domain expertise. In Descartes' case, examples of knowledge, other than immediately indubitable ones, are the result of his Method of Doubt (Descartes, 1637). Third, as we will argue, both programs, in their focus on the critical-rational status of knowledge, fail to properly account for or give credit to the important role of the intuitive-emotional side of knowledge in reasoning and knowledge management.

A KBS is a Cartesian creature if ever there was one, and it shares in the strengths and weaknesses of Descartes' original analysis of knowledge. We now turn to a weakness in Descartes' program, and a "virus" that may infect and seriously debilitate its KBS progeny. Damasio (1994) argues that Descartes' analysis of what is required to have and act on knowledge is in error; that there is an intuitive-emotional side to knowledge, which, if missing, will impede the reasoning process. Descartes' error is his "abyssal separation between body and mind" (p. 249), which, subsequently, does not allow for the role of the body-based side of knowledge in reasoning. In Damasio's analysis, what we understand as cognition is a collaboration of mind and body, not as Descartes would have it as a product of mind alone, and it is the embodied feelings that motivate individuals to reason and to act on their knowledge.

Jerome Bruner (1986), a pioneer in the cognitive psychology movement, suggested that the over-emphasis on disembodied thought had left a gap in our understanding of how human beings process information. The identification of thought with rule-governed reason was a biased way to regard knowledge:

> Take first the concept of thought. It is, to begin with a highly refined abstraction, an abstraction originally formulated in philosophy precisely to contrast it with activity governed by unreason and "tainted by passion." The defining character of thought is its product: the outcome of pure thought always passed the test of right reason. What did not conform was not, in the strict sense, pure thought. It was no accident that the mathematician George Boole entitled his famous work on algebra *The Laws of Thought*... If this Classical Abstraction had worked, the intersect between thought and emotion would constitute a null class. I am not intending only a logical joke, for it was certainly the hope of early logicians and philosophers to find some way of sorting out the chaff of unreason from the wheat of reason. And this was to be accomplished by the provision of finer and finer rules of right reason (that is, laws of logic) rather than by closer and closer descriptions of the activity of thinking itself (or, for that matter, of emotion) (pp. 106–107).

It is this "closer description" of thinking that is needed in order to understand how all users, from novice to expert, will respond to our attempts to package knowledge in a KBS.

THE WAYS OF THE FISH: A CLOSER DESCRIPTION OF THINKING

In attempting to answer the question, "What makes a user receptive to the knowledge in a KBS?," we would do well to be guided by the following metaphor: Well-known psychiatrist Milton Erickson regularly advised his students that for the successful fisherman, it is not the quality of the fishing equipment that counts, but the knowledge of the ways of the fish. Since the beginning of the Artificial Intelligence movement, researchers in this area have tended to disregard the ways of the fish in favor of the ways of the equipment. The assumption has been that people, when thinking at their best, think only logically.

Antonio Damasio (1994), a neuroscientist at the University of Iowa, has discovered that both reason and feeling must work in harmony in order for people to make good decisions. Damasio's findings are the result of studying patients with a particular type of brain injury. Damage to the frontal cortex results in a lack of ability to make good social judgments, while at the same time, such brain injured patients appear to retain most other intellectual abilities. The prototype of this sort of brain insult was the case of Phineas Gage, who in 1848 suffered massive damage to his frontal lobe area. Gage's case is a famous one, known to every student of psychology because of the unexpected alteration of his behavior following a particularly bizarre accident. Phineas Gage was a railroad foreman supervising blasting operations for the laying of new track near Cavendish, Vermont. Damasio's account is vivid:

> It is four-thirty on this hot afternoon. Gage has just put powder and fuse in a hole and told the man who is helping him to cover it with sand. Someone calls from behind, and Gage looks away, over his right shoulder, for only an instant. Distracted, and before his man has poured the sand in, Gage begins tamping the powder directly with the iron bar. In no time he strikes fire in the rock, and the charge blows up in his face. The explosion is so brutal that the entire gang freezes on their feet. It takes a few seconds to piece together what is going on. The bang is unusual, and the rock is intact. Also unusual is the whistling sound, as of a rocket hurled at the sky. But this is more than fireworks. It is assault and battery. The iron enters Gage's left cheek, pierces the base of the skull, traverses the front of his brain, and exits at high speed through the top of the head. The rod has landed more than a hundred feet away, covered with blood and brains (p. 4).

Two surprises were in store for those who followed the case of Phineas Gage: first, he did not die, or even become unconscious, as the result of his injury. He was carried into town on the bed of a wagon, and was able to talk and even walk with assistance. Second, and even more remarkable, despite major insult to large areas of his frontal lobes, he did not lose to any great extent his "high-reasoning" skills.

The iron bar which passed through Gage's brain weighed 13.25 pounds, was 3 ft. 7 in. in length, and was 1.25 in. in diameter (Damasio, 1994). The attending physician, John Harlow, left a detailed account of Gage's subsequent recovery. According to Dr. Harlow, Gage regained his former strength, was not paralyzed, had no noticeable difficulty with speech or language, and his only impairment at first seemed to be blindness in his left eye. But later it became apparent that something even more important had been lost, something Harlow described as the "equilibrium, or balance, so to speak, between his intellectual faculty and his animal propensities" (cited in Damasio, 1994, p. 8). Gage was now fitful, irreverent, profane in the presence of ladies, seemingly indifferent to other people, impatient with advice going contrary to his wishes, possessing no planning ability, and unable to sustain effort. He seemed to have little impulse control or social acumen.

Damasio has studied more that a dozen patients whom he calls "modern Gages" and who have sustained damage to nearly the same area of the frontal cortex. From these examples and other findings in neurology, Damasio has come to believe that feelings in the form of "somatic-markers" are indispensable for good judgment. Against conventional wisdom, which considers emotion and feelings to be atavistic and contaminants of better judgment, Damasio believes them to be a crucial part of good decision making in most human contexts.

It does not seem sensible to leave emotions and feelings out of any overall concept of mind. Yet, respectable scientific accounts of cognition do precisely that...[but] *feelings are just as cognitive as any other perceptual image*...they offer us *the cognition of our visceral and musculoskeletal body state* as it becomes affected by preorganized mechanisms and by cognitive structures we have developed under their influence. Feelings let us *mind the body*, attentively, as during an emotional state, or faintly, as during a background state...I see feelings as having a truly privileged status (pp. 158–159, emphasis in original).

Damasio's notion of the privileged status of emotions and feelings is not only that they came first in evolution, but that reasonable thinking only works well when supported by affective processes. Reduction in emotion renders rational thought incapable of dealing with our most important problems.

When discussing what he identifies as Descartes' error, viz., "the abyssal separation between body and mind...," and explaining how this particular error is related to Descartes' famous "Cogito ergo sum," Damasio (1994) wrote that, "Taken literally, the statement illustrates precisely the opposite of what I believe to be true about the origins of the mind and body. It suggests that thinking, and awareness of thinking, are the real substrates of being" (p. 248).

Damasio argues against the long-standing prejudice that placed thought in a sanctified location "in the head," while all corruptors of pure reason existed from the neck down. "Upstairs in the cortex there is reason and willpower, while downstairs...there is emotion and all that weak, fleshy stuff" (p.128). Rather than being merely the life support for the brain, the body — with its attendant emotions and feelings — provides the real basis for rational thought.

> The apparatus of rationality, traditionally presumed to be *neo*cortical, does not seem to work without that of biological regulation, traditionally presumed to be *sub*cortical. Nature appears to have built the apparatus of rationality not just on top of the apparatus of biological regulation, but also *from* it and with it. The mechanisms for behavior beyond drives and instincts use, I believe, both the upstairs and the downstairs: the neocortex becomes engaged *along with* the older brain core, and rationality results from their concerted activity (p. 128-emphasis in original).

This then calls for a closer description of thinking, in order to better understand how both expert and novice users will respond to our attempts to package knowledge in a KBS. On the neurophysiological level, a closer description of "thinking" establishes two points: (1) Descartes' argument for the separation of body and mind and that all reasoning is independent of the body is untenable; and, (2) equally important, the traditional brain-based analysis of rationality as primarily neocortical also needs revision, to account for the substantial role of the subcortical brain and other body parts in providing the requisite intuitive-emotional support for rational thought.

Support for our analysis of the role of the intuitive-emotional side of knowledge in reasoning is also provided by computer scientists who are attempting to get machines to be creative. A recent example, complementary to Damasio's argument, is the work of David Gelernter, a computer scientist at Yale. Gelernter (1994) calls for a new folk psychology which would include the emotions, intuitions, and images that support creative thinking. He begins by stating that..."adding 'emotions' to computers is the key to the biggest unsolved intellectual puzzle of our time: how thinking works" (p. 2). Gelernter argues for a spectrum of human thought ranging from highly focused to loosely focused. The logical side of thought is highly focused..."Almost all attempts to simulate thought on a computer have dealt exclusively with this narrow, high-focus band at the top of the spectrum" (p. 5). Gelernter defines an emotion as the "glue of thought" when focus is loose; "a mental state with physical correlates; it is a felt state of mind..." (p. 27). He, like Damasio, believes that an affective link exists in all complete knowledge. When we only emphasize the abstract logical side of knowledge, then we do not experience the satisfaction of truly possessing it. So, how do we glue a user to a KBS embodiment of knowledge, and how must the knowledge be embodied?

KNOWLEDGE AS CRITICAL AND INTUITIVE KNOWING: THE BELIEF MATRIX

Howard Margolis (1987), a cognitive scientist at the University of Chicago, has offered a model that we consider to be useful in considering the problem of modeling the other side of knowledge and its connection to the critical-rational side of knowledge. Between the contexts in which we live and the concepts we form about our experiences, there exists a dialectical tension. Margolis (1987) calls this tension "cognitive rivalry." Whenever we are faced with an argument as to the truth of something, e.g., an argument of reasoning leading to a decision (automated or not), the outcome of this rivalry will leave us with a particular "affective state" concerning the argument.

Margolis' cognitive model, which he refers to as the Belief Matrix, represents the way in which we come to regard the products of these opposing processes as knowledge. He suggests that logical and factual arguments are submitted to a critical scrutiny as to whether there is sufficient reason to believe their conclusions. In other words, the critical criterion (the "C-question") is "Do the reasons look convincing?" Statements such as: "I've got a good reason" or "Logically, it has to be that way" are common responses to this question. This reasoned approach results in "knowing about" the world. The rival to the critical argument is intuition. Margolis (1987) stated that it is not enough to satisfy the "C-question," but it is also the "I-question:" "Does the result look right?," which must be answered. We would understand variations on Margolis' I-question to include: "Does the result feel right?" Statements such as "I've got a hunch" or "I just know" or "…works for me!" are more typical responses to these questions at the level of intuition and feeling.

In Margolis' model, the outcome of the rivalry between critical and intuitive questioning of the argument will create one of nine affective states. When asking the question, "Do the reasons look convincing?," one may respond with a "yes" (+), or a "no" (–), or a "maybe" (o). Likewise, an answer to the question, "Does the result look (or feel) right?," can be "yes" (+), "no" (–), or "maybe" (o). If one answers "yes" to the "C-question," it could be said that the person is convinced by the argument. In other words, a critical analysis of the argument by this person finds it to be free of logical flaws. If the person answers "yes" to the "I-question," this might be considered an indication that the person is satisfied by the argument. In other words, the rechecking of intuition has not given indication of any distortion or discomfort brought about by a conflict with one's personal acquaintance with such things.

Thus, Margolis' model is a 3×3 matrix (Figure 12.1) forming nine "affective states" which represent the possible outcomes when critical cognition converges with intuitive cognition. However, only one of the affective states is *knowledge*. Knowledge is the result of agreement between positive (+) critical and a positive (+) intuitive judgments: a positive critical judgment is a judgment based on an argument's reasons looking convincing, that is, on the appearance of validity; a positive intuitive judgment is based on the argument's results looking/feeling right, that is, on the appearance (experience) of comfort (Margolis, 1987). In order for someone to say that an argument's reasons look convincing, an image of former experience of validity would have to be summoned for comparison. For someone to say that results looked/felt right, an emotional image of comfort would have to be recalled.

As just stated, for someone to experience the affective state of knowledge, critical and intuitive processes must both confirm the argument. But what if they don't? Then we experience a different affective condition than knowledge. In the extreme of such cases — in which one process (C or I) answers "yes" and the other answers "no" — we are in a state of paradox. The cell in the upper right corner of the matrix (C–,I+), might be called (the paradox of) "Faith." For a person to believe something to be true, without that person being able to muster a critical argument in its behalf, is an act of fidelity to something which they themselves cannot support critically. Religious devotion, for example, is something which goes beyond mere Belief

C+ / I+	Co / I+	C- / I+
Knowledge	**Belief**	**Paradox**
C+ / Io	Co / Io	C- / I o
Doubt	**Uncertainty**	**Doubt**
C+ / I-	Co / I-	C- / I-
Paradox	**Disbelief**	**Contrary Knowledge**

FIGURE 12.1.   The belief matrix. (Adapted from Margolis, 1987. With permission.)

(Co,I+); it is a certainty of belief in the face of convincing contrary evidence. Most people are well acquainted with the experience of faith, and they often appear to find this affective state to be satisfying and comfortable. If faith is challenged by reason, one can defend against this assail by refusing "to be confused by the facts," by rejecting the basis for the argument, or by simply withdrawing from the arena in which such reasoning is taking place. This is why so many religious groups create communities with strong boundaries that keep out the ideas of unbelievers.

The paradox represented by the lower left cell of the matrix (C+,Io), appears to be what scientists smugly call the "counterintuitive." A good example of this affective state is illustrated in the "Super String Theory" argument. Since the time of the Greeks, humans have grown used to the idea that the universe is made up of atomic particles: the tiny building blocks of everything. Like some great pointillist painting by Seurat, the entire panorama was considered to be wrought of infinitesimal dots of stuff so tightly held together, that it looked (was experienced as) solid. Now we are told, by Edward Witten (cited in Cole, 1987), that it is not particles that make up everything, but rather strings: tiny loops which vibrate invisibly and may possess as many as ten dimensions. There are the regular dimensions that we know: height, breadth, and depth; plus time. In addition, there are six hidden dimensions. The vibrating string, with its ten dimensions, resonates in many modes and it is the vibrations which result in all the particles and forces in the universe. This theory appears to have developed some currency in physics and may serve to resolve many of the current paradoxes brought about by quantum mechanics.

If this is the first time you have come across string theory, you probably greet this argument with some degree of dissatisfaction. But if you can imagine yourself

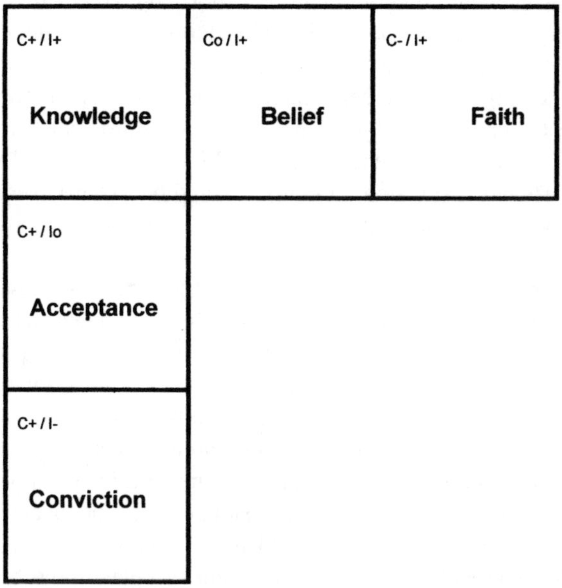

C+ / I+	Co / I+	C- / I+
Knowledge	**Belief**	**Faith**

C+ / Io
Acceptance

C+ / I-
Conviction

FIGURE 12.2. The judgment matrix. (Adapted from Presbury and Benson, 1994, and Margolis, 1987. With permission.)

being compelled by the facts and becoming convinced by the argument of string theory proponents, you would likely be experiencing the affective state of counter-intuitive paradox (C+,I–) as you struggle for an intuitive grasp of the concept, in the hope of achieving balance. To be convinced by an argument, however, is not the same as being satisfied by it. The lingering dissatisfaction that may come with holding a convincing idea without having achieved intuitive balance, might be managed by defense mechanisms: detaching from, or depersonalizing, the argument. In the case of the KBS user, this would be a problem.

THE JUDGMENT MATRIX

We have found Margolis' (1987) Belief Matrix very helpful in our attempts to make sense of the relationship of critical and intuitive thinking. However, to further our project, it needs modification. First, remember, our interest is in the role of the critical-rational and intuitive-affective sides of knowledge in KBS that reason to conclusions, e.g., judgments, recommendations, or decisions. Accordingly, we have modified the name to the "Judgment Matrix." Second, we are only interested in those cells where at least one of the questions (the critical-question or the intuitive-question) is answered affirmatively (+), because only cells where either the C-question or I-question is answered affirmatively could be used to justify representing and including a belief in a KBS or would tempt someone to use the KBS' output. Figure 12.2 shows the Judgment Matrix with its five cells of conditions for knowledge. Finally, we labeled Margolis' cells to reflect our argument. The cells where at least one of the knowing conditions (sides of knowledge) is positive (+) are labeled: Knowledge (C+,I+); Acceptance (C+,Io); Conviction (C+,I–); Belief (Co,I+); and Faith (C–,I+).

The only knowing condition cell which does not reflect some form of cognitive dissonance (Festinger, 1957) is knowledge. Every condition cell other than knowledge would, therefore, be challenged by what Margolis called an "affective state" of uncertainty. The ideal situation, of course, would be to have everything that one "knows" exist within the knowing condition of knowledge. For most of us, this is not the case.

We further conjecture that if someone has made judgments based upon any of the cells except knowledge (as we all have), then these judgments have been to some extent unstable. There would appear to be, in all of us, a drive to resolve this instability. We seek reasons for our intuitive leaps of faith through after-the-fact rationalizations. Likewise, we attempt to find intuitive comfort in situations where we have been convinced of an initially counterintuitive idea. For example, most of us have grown comfortable with the idea of living in a heliocentric solar system, despite the fact that we cannot sense this situation except in drawings or those mechanical models we were shown in elementary school.

The knowledge condition is comfortable, and acceptance and belief are conditions that are probably not extremely uncomfortable, since they are only challenged by a "maybe" (+,o). But the knowing conditions of conviction and faith, the paradox conditions, present unsettling dissonance. Each of these is assailed by strong opposition from its opposite mode. In conviction, the critical question is affirmed, while the intuitive question is rejected. In faith, the intuitive question is affirmed and the critical question is rejected. These paradox conditions (+,–) would be affectively intolerable if not defended in some way. Knowledge engineers must attempt to assist users of KBS to come as close to Margolis' knowledge cell as possible. Since there is a drive in all people to reduce the dissonance (Festinger, 1957), we hypothesize that this applies to the knowing — thus a drive to arrive at knowledge. We would not wish to push our user into a position of defense. Therefore, all critical-rational representations of knowledge must be presented in such a way as to invite a confluence with intuition-emotion. Any "knowledge" offered by the computer must be tailored for a fit with the user's experience of the world.

EMOTIONS AND THE OTHER SIDE OF KNOWLEDGE

As noted previously, the professional KBS practitioner's promise is to deliver sound and valuable KBS products. In the past, soundness has been seen as a critical-rational justification problem while value-as-used has been addressed as a "friendly interface" issue. We are suggesting that there is more to insuring a KBS is valuable than might be captured with a "friendly" user interface. Insuring that a KBS is valuable requires that the system be used and this, in turn, will form a user's positive emotional response to the system's output, e.g., be it a description of the knowledge-base, a recommendation from its inference engine, or an explanation of its behavior. These positive emotional responses associated with the systems output are required if the user is to act on the system's output. So, the question becomes, how do we insure a positive intuitive-emotional response to the output?

Oatley and Jenkins (1996) have surveyed the vast literature on emotions and have, they say, "rashly" attempted the following definition of an emotion:

1. An emotion is usually caused by a person consciously or unconsciously evaluating an event as relevant to a concern (a goal) that is important; the emotion is felt as positive when a concern is advanced and negative when a concern is impeded.
2. The core of an emotion is a readiness to act and the prompting of plans; an emotion gives priority for one or a few kinds of action to which it gives a sense of urgency — so it can interrupt, or compete with, alternative mental processes or actions. Different types of readiness create different outline relationships with others.
3. An emotion is usually experienced as a distinctive type of mental state, sometimes accompanied or followed by bodily changes, expressions, actions (Jenkins p. 96).

If we accept this definition, eliciting an emotional response requires eliciting an evaluation, i.e., assigning value to some thing or event. Given our analysis, eliciting a positive emotional response means a user sees the critical-rational side (presentation) of the KBS output and experiences its intuitive-affective component as positive (+). That is, a user is in a (C?,I+) cell of the judgment matrix. That the user's experience is correctly represented by this cell establishes the output as "meaningful" for the user. As per Damasio, the explanation for this phenomenon is that the user recognizes the systems output as "like" a past attempted solution to the problem and this past solution is associated with a positive intuitive-affect. Eliciting an emotional response also leads to a readiness (disposition) to act or prompting of plans. In the case of a positive emotional response to the system's output, this amounts to acting (or being disposed to act) on the KBS output/recommendation. Finally, if this is all to take place, there must be associated distinctive mental and bodily states.

COUNSELING AND KNOWLEDGE ENGINEERING: ASKING AND TELLING

Counseling and knowledge engineering have much in common: both hope to instill in (make available to) the client (the user) declarative and procedural knowledge that will make that person more fully functional in some domain. Broadly understood, "knowledge acquisition" refers to all information gathering activity in the problem identification, engineering, development, delivery, maintenance, and user training associated with a KBS project. Knowledge acquisition is not unique to KBS development; it exists in every human communication transaction. Professional knowledge engineers might profit from some of the hard-won insights in the counseling field.

"Why can't you just tell'em?" This is the question often asked by beginning students of counseling techniques. If a client has a problem and the counselor believes he or she has the answer, then novice counselors think the most efficient thing to do would be to "just tell'em." The only problem with this strategy is that it usually does not work. A similar assumption is present in the KBS design: ideally, the knowledge from the expert(s) in a domain will be extracted, engineered, and delivered to a user, and, as a result, the user would then function at an improved level. Knowledge engineers know that this expectation often falls short. For example, a knowledge engineer may be more or less studied (street smart) in interview theory, preparation, and techniques; more or less skilled in managing an interview session; and more or less disposed to this type of work. Likewise, domain experts may be more or less articulate about their knowledge, uncomfortable participating in a KBS project, unwilling to give up their expertise, or irritated by the manner in which the knowledge engineer puts the questions. Less than optimal interaction here will result in the well-known knowledge acquisition "bottleneck" of knowledge engineering (Luger and Stubblefield, 1993). The success of any given knowledge acquisition session depends on how all these parameters play out.

At the other end of the process (this is where "you tell'em") lurks what might come to be called the knowledge reception bottleneck problem of counseling. A cardinal principal in the counseling profession is that knowledge is not "knowledge" until the client *owns* it. Ownership occurs when the client either assimilates or accommodates the information so that it fits into his or her internal cognitive scheme. Information remains irrelevant, advice remains gratuitous, and gems of wisdom go unheeded until they fit into the client's way of representing reality. This is the pedagogical problem: how to package information so that people will truly learn it, i.e., make it their own, and put it to use in their lives, be they a counseling client, a KBS user, or a student.

When new information is presented to anyone, it is experienced as strange because that person may not have a representational scheme into which the knowledge can be assimilated. This means that the knowledge representation of the client/user/learner must be changed. People are reluctant to do this; they would rather assimilate new knowledge into their existing categories. Accommodating to new knowledge is an uncomfortable process.

> Briefly, assimilation involves making a response that has already been acquired; accommodation is the modification of a response. Or, to put it another way, to assimilate is to respond in terms of preexisting information; it often involves ignoring some aspects of the situation in order to make it conform to aspects of the...mental system. In contrast, to accommodate is to respond to external characteristics; as a result, it involves making changes in the...mental system (Lefrancois, 1988, p. 181).

Because of the discomfort involved in accommodating to new information, people will resist putting themselves through the experience. Even in counseling situations, in which the majority of clients come seeking change, it has been estimated that 50–75% of them could be described as reluctant (Young, 1992). Will the same hold true for KBS?

Jerome Frank has, over many years (1971, 1981, 1991), conducted a meta-analysis of the common factors that create effectiveness in counseling situations. After a lifetime of analysis, he reduced his list to six mechanisms of change:

1. The strength of the therapist/client relationship
2. Methods that increase motivation and expectations of help
3. Enhancing a sense of mastery or self-efficacy
4. Providing new learning experiences
5. Arousing emotions
6. Providing opportunities to practice new behavior (cited in Young, 1992, p. 15).

It is well known that, in a knowledge engineering project, pitfalls and bottlenecks have much more to do with human factors than with programming. The knowledge engineer who is successful employs the above factors, just as a counselor would, in order to enhance the possibilities of eliciting accurate and complete information from the expert(s) and ensuring that the information will be useful and used by the client(s). Effective counselors know that the quality of the relationship established with clients is what sustains all other factors, and that the timing of the other five factors is crucial for complete success. Timing comes from "reading" the client and backing off when the client displays resistance to what is happening in their relationship. A counseling session begins when the counselor attempts to elicit information from the client. This is not unlike the knowledge elicitation process employed by the knowledge engineer with an expert, where the expert is also the anticipated user. In either case, care must be taken so that the expert/client does not feel that he or she is being "raped" or "strip-mined" of the information.

It is a little-appreciated fact that when someone is asked to make unconscious or automated knowledge conscious and explicit, there is an almost "obscene" quality to the experience. For example, if one were asked to spit into a glass until it filled, and then were asked to drink it, it is likely the person would feel disgusted by the idea. Yet we are doing this all the time when we swallow our own sputum. It is the making of an intuitively known fact explicit that often renders it uncomfortable. It changes the context of the knowledge and alters its meaning. In addition, anyone who does something well, in an almost "mindless manner," is sometimes reluctant to analyze it for fear of interfering with its smooth operation. Poets and graphic artists typically express this reticence, and everyone knows that too much self-consciousness applied to doing anything that has been previously automated often leads to an awkward performance.

The conditions that the knowledge engineer would wish to create for an expert who is not the target user, based on Frank's factors, would be: (1) a trusting relationship, (2) the belief on the part of the expert that elicitation is not exploitation, and that he or she will be helped — rather than diminished — by the process, (3) the feeling that the expert is learning more about his or her own capabilities by making them more explicit, (4) the experience of knowing "afresh" (clients in counseling often report after explicating their knowledge, that they paradoxically "knew, but didn't know it all along"), (5) the arousal of positive affect and investment in the project, and (6) giving the expert an experience of learning about knowledge engineering and the authority for confirming whether the engineer has organized the information properly.

The conditions that the knowledge engineer would wish to create for a user who is not the domain expert, based on Frank's factors, would be: (1) a trusting relationship, (2) the belief on the part of the user that he or she will be helped — rather than diminished — through using the system, (3) the feeling that the user is learning more about the domain and increasing his or her skills, (4) the experience of verifying knowledge about the domain they "sort-of thought they had," (5) the arousal of positive affect and investment in the project, and (6) giving the user an experience of learning about knowledge engineering and the authority for confirming whether the engineer has presented the information properly, that is, in a useful way.

Both counseling-as-knowledge-engineering and knowledge-engineering-as-counseling need project buy-in, continued project support, and proud ownership of the final product, on the part of all project participants. In the special cases of domain experts and users, this buy-in, continued support, and ownership is required. Implementing these conditions will help realize these needs and requirements.

THE DASEIN DESIGN

We have all had the experience of listening to someone tell of an experience that was, in some way, meaningful to that person — but we did not get it. It might be about a funny happening or a moment in which the person was profoundly moved. The story-teller, noting our blank expression after attempting to insert additional information so that we will finally understand, ultimately aborts the communication with an exasperated, "forget it...you had to be there!"

Being there was the central notion of Heidegger's (1962) existential philosophy. He repeatedly distinguished between the "truth of Being" and the "truth of propositions." The former was a type of lived experience, while the latter was formal, mediated, vicarious experience. The first wave of KBS development was predicated on the truth of propositions: the critical-rational side of knowledge. The next wave, we propose, will add "being there" to the equation: the intuitive-emotional experience of the user will then serve as a complement to the Cartesian side. Heidegger's term for being there was the German word "Dasein" (Da' zign). In order to improve our knowledge-based systems, we need to add a Dasein design. Just how this is to be done is still unclear, but we believe we have offered a compelling argument for the necessity of such a move.

What would the components of such a system be? Perhaps the notion of "being there" offers some clues. Had we been there when our story teller's event actually took place, we would have had "knowledge by acquaintance" and some first-hand (intuitive-emotional) experience would have been added that was not present in the telling: First of all, the event would have been real, immediate, and concrete, rather than something to be constructed from the story teller's account. Secondly, it is possible that we would have been "moved" in a way similar to how our story teller was moved. In other words, some emotional stirring might have taken place within us. Finally, the experience would have been ours, rather than being a mediated account belonging only to the story teller. We would have been participants in the event; and so it would not have been a vicarious happening, but one in which we were present.

In the communication transaction described above, the story teller attempted to make us see — to make us feel. The story teller was the knowledge engineer and we were the user of the story. Did our understanding come up short because that's just the way things are…you had to be there? Or was there something missing in the telling? Obviously, mediated experience always loses something in the telling.

But what takes place when we are moved by a passage of literature or a scene in a movie; or when we are drawn into an experience by any type of media? Our answer would be that what has happened in such a case is that the Dasein design of the medium has fit neatly into our experience so that it seems as if we are there. Authors, movie directors, and graphic artists are knowledge engineers. If they successfully engineer their communication, we get it. We see the truth of it. This is not the truth of propositions, but the truth of Being. A good story teller can create the Dasein experience in the audience (users).

The story teller's trick appears to be a matter of creating intuitive-emotional images to which all listeners or viewers can relate. For some time in the A.I. field, researchers have been working on how to create a good story. For example, this emphasis on the narrative approach is evident in Roger Schank's attempts to create story-telling programs, the development of frames (Crevier, 1993), and the analogy projects of Gelernter (1994) and Hofstadter (1995).

Knowledge engineers who wish to engage the "other side" of knowledge — the topic of this chapter — will do well to consider a Dasein design. The elements of such a KBS system are, at present, not completely known. But, apparently, there are two salient questions to continually ask when getting either expert or user feedback on a KBS: (1) "Does it look right?," and, of equal importance, (2) "Does it feel right?" We have attempted to make the problem as clear as we can and to emphasize its importance. We expect that minds more creative than ours will take up the challenge of this problem and solve it.

REFERENCES

1. Bruner, J., *Actual Minds, Possible Worlds.* Cambridge, MA: Harvard University Press, 1986.
2. Cole, K.C., A theory of everything. *New York Times Magazine*, October 18, 20–28, 1987.
3. Crevier, D., *AI: The Tumultuous History of the Search for Artificial Intelligence.* New York: Basic Books, 1993.
4. Damasio, A., *Descartes' Error: Emotion, Reason, and the Human Brain.* New York: G.P. Putnam's Sons, 1994.
5. Descartes, R., Discourse on method (Part II) In J. Cottingham, R. Stoothoff, and D. Murdoch (Trans.) 1637 *The Philosophical Writings of Descartes* (Vol. 1). Bath: England: Cambridge University Press p. 120, 1985.
6. Frank, J.D., Psychotherapists need theories. *Int. J. Psych.*, 9, 146–149, 1971.

7. Frank, J.D., Therapeutic components shared by all psychotherapies. In J.H. Harvey and M.M. Parks (Eds.) *Psychotherapy Research and Behavior Change*. Washington, DC: American Psychological Association, pp. 175–182, 1981.

8. Frank, J.D. and Frank, J.B., *Persuasion and Healing* (3rd ed.). Baltimore: Johns Hopkins University Press, 1991.

9. Festinger, L., *A Theory of Cognitive Dissonance*. Palo Alto, CA.: Stanford University Press, 1957.

10. Freedman, D., *Brainmakers: How Scientists are Moving Beyond Computers to Create a Rival to the Human Brain*. New York: Simon & Schuster, 1994.

11. Gelernter, D., *The Muse in the Machine: Computerizing the Poetry of Human Thought*. New York: The Free Press, 1994.

12. Heidegger, M., *Being and Time* (J. Macquarrie and E.S. Robinson, Trans.). New York: Harper & Row, 1962.

13. Hofstadter, D., *Fluid Concepts and Creative Analogies: Computer Models of the Fundamental Mechanisms of Thought*. New York: Basic Books, 1995.

14. Lefrancois, G.R., *Psychology for Teaching*. Belmont, CA: Wadsworth, 1988.

15. Luger, G.F. and Stubblefield, W.A., *Artificial Intelligence: Structures and Strategies for Complex Problem Solving* (2nd ed.). Redwood City, CA: Benjamin/Cummings, 1993.

16. Margolis, H., *Patterns, Thinking, and Cognition: A Theory of Judgment*. Chicago: The University of Chicago Press, 1987.

17. Newell, A. and Simon, Computer science as empirical inquiry: Symbols and search (The Tenth Turing Lecture), 1976. Reprinted in M.A. Boden (Ed.) *The Philosophy of Artificial Intelligence*. New York: Oxford University Press, 1990.

18. Oatley, K. and Jenkins, J.M., *Understanding Emotions*. Cambridge, MA: Blackwell, 1996.

19. Presbury, J.H., Benson, A.J., and McKee, J.E., The mock turtle's lament: The cost of critical thinking. *Virginia Counselor's Journal*, Vol. 20, Spring, 12–19, 1992.

20. Young, M.E., *Counseling Methods and Techniques: An Eclectic Approach*. New York: Merrill, 1992.

Index